# Nonlinear Analysis

# Nonlinear Analysis

Sudarsan Nanda

**Alpha Science International Ltd.**
Oxford, U.K.

**Nonlinear Analysis**

188 pgs.

**Sudarsan Nanda**
Professor of Eminence and Research Chair
KIIT University
Bhubaneswar

Copyright © 2013

ALPHA SCIENCE INTERNATIONAL LTD.
7200 The Quorum, Oxford Business Park North
Garsington Road, Oxford OX4 2JZ, U.K.

**www.alphasci.com**



ISBN 978-1-84265-732-4

Printed in India

**To**
My wife, daughters,
teachers and students.

# Preface

In this monograph we present the developments in the areas of Nonlinear Analysis. The topics covered include calculus in Banach space, Convex sets, Convex and Invex Functions, Best Approximation, Fixed Point Theorem, Nonlinear operators, Variational Inequality, Complementarity Problem and Semi-inner-product space. Some of the results presented here did not appear in any book before. This book will serve the purpose of a course on Nonlinear Functional Analysis which is a second course on Functional Analysis for MSc/Mphil students and prerequisite is a first course on Functional Analysis. The manuscript will be useful for researchers working in this field and it could be used as a one-semester course for graduate students and research scholars. The purpose will be best served if it benefits those for whom it is meant.

Although linearity plays an important role in mathematics and applications, nonlinearity is still more important and useful since many real world problems are nonlinear (even nonconvex) and also nonsmooth in nature. Hence a rich theory has been developed in Nonlinear Functional Analysis and in Nonconvex Analysis. Fixed Point Theory dates back to Banach in early 1920s whereas the theory of nonlinear operators was developed during early 1960s. Variational Inequality and Complementarity Problem being originated in 1960s by Stampacchia, Lions, Cottle and others have been developed to a great extent in theory, algorithms and applications.

Materials have been taken from various authors and the sources have been recognized by mentioning those in the bibliography at the end. Enough care has been taken for presentation and for making the book free of errors. Despite our efforts mistakes might be there. The author would be grateful if any error or apparently better approach is pointed out.

The author wishes to thank Indian Institute of Technology, Kharagpur and KIIT University, Bhubaneswar for making facilities available for preparation of the manuscript.

**Sudarsan Nanda**

# Contents

# Calculus in Banach Space

## INTRODUCTION

In this chapter we discuss calculus in real Banach spaces. In Section 1.1 we deal with Gateaux and Frechet derivatives of operators, not necessarily linear, between Banach spaces, develop Mean Value Theorems, discuss relationship between the two types of derivatives. In Section 1.2 we discuss the concept of integration.

## 1.1  GATEAUX AND FRECHET DERIVATIVES

Let $X$ and $Y$ be real Banach spaces.

**Definition 1.1.1**  An operator $F:X \to Y$ is said to be Gateaux differentiable at a point $x \in X$ iff there exists a continuous linear operator $A:X \to Y$ such that

$$\lim_{t \to 0} \frac{F(x + th) - F(x)}{t} = Ah$$

for every $h \in X: A$ is called the Gateaux derivative of $F$ at $x$ and its value at $h$ is denoted by $\delta F(x, h)$. If $F$ is Gateaux differentiable at every $x$, we write $F(x)$ for $\delta F'(x, h)$.

**Example 1.1.1**  Consider the Hammerstein operator $F:L_p [0, 1] \to L_p [0, 1]$ given by

$$[F(x)] (t) = \int_0^1 k(t, s) f(s, x(s)) \, ds$$

where $k$ and $f$ are such that the derivative can be taken under the integral sign.

By definition, $\qquad \delta F(x, h)(t) \left[ \frac{d}{d\tau} F(x + \tau R)\big|_{\tau=0} \right](t)$

$$= \left[\frac{d}{d\tau}\left[\int_0^1 k(t,s)\,f(s,(x+\tau h)(s)ds)\right]\right]_{t=0}(s)$$

$$= \left[\int_0^1 k(t,s)\frac{d}{d\tau}\,f(s,(x+\tau h(s)ds)\right]_{t=0}(s)$$

$$= \left[\int_0^1 k(t,s)\,f'_x(s,x(s)h(s))\,h(s)\,ds\,(s)\right]$$

That is, the Gateaux derivative $F'(x)$ of the Hammerstein operator is the linear integral operator with the Kernel $K(t,s)\,f'_x(s,x(s))$ and is given by the above equation.

We note that an operator from $X$ to $R$ is called a functional. For a functional $f_1$ $\delta f(x,h)$ is called the gradient of $f$ and is also denoted by $\nabla f$. Thus $\nabla f$ is a mapping from $X$ to the dual $X^*$ of $X$.

We first present a mean-value theorem for functionals.

### 1.1.1 Mean Value Theorem for Functionals

**Theorem 1.1.1**    Let $f:x$ have Gateaux derivative $\delta f(x,h)$ at every point $x \in X$. Then for the points $x, x+h \in X$, there exists a constant $\theta$, $0 < \theta < 1$,

$$f(x+h) - f(x) = \delta f(x+\theta h, h).$$

**Proof.**    Define a function $\varphi$ by $\varphi(t) = f(x+th)$.

Then
$$\varphi'(t) = \lim_{s \to 0}\left[\frac{\varphi(t+s) - \varphi(t)}{s}\right]$$

$$= \lim_{s \to 0}\left[\frac{\varphi(x+th+hs) - f(x+th)}{s}\right]$$

$$= \delta f(x+th, h)$$

Now,    $\varphi(1) - \varphi(0) = f(x+h) - f(x)$

Applying Mean Value Theorem for scalar functions of differential calculus we get

$$\varphi(1) - \varphi(0) = \varphi'(\theta),\ 0 < \theta < 1$$

Hence

$$f(x+h) - f(x) = \delta f(x+\theta h, h)\ \text{and this completes the proof.}$$

## 1.1.2  Variant of Mean Value Theorem for Operators

**Theorem 1.1.2**  Let $F: X \rightarrow Y$ have Gateaux derivative $F'(x)$ at every $x \in X$. Then for points $x$, $x + h \in X$ and $e \in Y^*$, $\exists$ a constant $\theta$, $0 < \theta < 1$ such that

$$(e, F(x + h) - F(x)) = (e, \delta F(x + \theta h, h)).$$

Further $F$ satisfies the Lipschitz condition

$$\|F(x + h) - F(x)\| \leq \| \delta F(x + \theta h\| \|h\|$$

**Proof.**  For $e \in Y^*$, define a functional $\varphi$ by

$$\varphi(x) = (e, F(x)).$$

Then

$$\frac{\varphi(x + th) - \varphi(x)}{t} = \left( e, \frac{F(x + th) - F(x)}{t} \right)$$

Taking the limit as $t \rightarrow 0$ and using the continuity of inner product we get

$$\delta\varphi(x, h) = (e, \delta F(x, h)).$$

Since Mean Value Theorem is valid for functional $\varphi$ on $X$ we get

$$\varphi(x + h) - \varphi(x) = \delta\varphi(x + \theta), \quad 0 < \theta < 1.$$

For the second part, since $e$ is arbitrary, by Hahn – Banach Theorem we can choose $e$ of unit norm such that

$$\|(e, F(x + h) - F(x))\| = \|F(x + h) - F(x)\|$$

Now we get

$$\|F(x + h) - F(x)\| = |(e, F(x + h) - F(x))|$$

$$= |(e, \delta F(x + \theta h, h))|.$$

$$\leq \| \delta F(x + \theta h, h)\|$$

$$\leq \|\delta F(x + \theta h)\| \|h\|$$

We know that a function $f$

$R \rightarrow R$ has a derivative $m$ at a point $a \in R$ iff for each $\epsilon > 0$, $\exists \delta > 0$ such that

$$\left| \frac{f(x) - f(a)}{x - a} - m \right| \leq \epsilon \quad \text{for } |x - a| < \delta$$

i.e. $\qquad |f(x) - f(a) - m(x - a)| \leq \epsilon \, |x - a| \quad \text{for } |x - a| < \delta.$

**Definition 1.1.2**   Let $X$ and $Y$ be real nls. $A$ a nonempty open subset of $X$ and $f: A \to Y$.

$f$ is said to be Frechet differentiable at a point $a \in A$ iff $\exists$ a linear map $T: A \to Y$ such that for each $\in > 0$, $\exists \delta > 0$:

$$\|f(x) - f(a) - T(x - a)\| \leq \in \|x - a\|$$

or
$$\lim_{\|h\| \to 0} \frac{\|f(x + h) - f(x) - T(h)\|}{\|h\|} = 0$$

for all $x \in A$ with $\|x - a\| < \delta$.

[$T$ need not be bounded]

This is a generalization of the scalar case.

*Remark 1.1.1*   Norm is a continuous function, for if $X$ is a nls, for $x$, $y \in X$,

$$|\|x\| - \|y\|| \leq \|x - y\|.$$

**Definition 1.1.3**   $\| \ \|: X \to R$ is said to be Frechet differentiable at a point $a$ iff $\exists$ a linear functional $g: X \to R$ such that for each $\in > 0$, $\exists \ \delta > 0$:

$$|\|x\| - \|a\| - g(x - a)| \leq \in \|x - a\| \text{ for } \|x - a\| < \delta.$$

**Example 1.1.2**   Let $f: R^2 \to R$ be defined by

$$f(x_1, x_2) = \begin{cases} \dfrac{x_1 x_2^2}{\left(x_1^2 + x_2^4\right)} & \text{if } x_1 \neq 0 \text{ and } 0 \text{ if } x_1 = 0 \end{cases}$$

At the origin $(0, 0)$, $f$ is Gateaux differentiable but not continuous. For

$$\frac{f(a \ R)}{a} = \frac{a h_1 a^2 h_2^2}{a\left(a^2 h_1^2 + a^4 h_2^4\right)}$$

$$= \frac{h_1 h_2^2}{h_1^2 + a^2 h_2^2} \to \frac{h_2^2}{h_1} \text{ as } \alpha \to 0$$

In particular if $\quad h = (h_1, 0)$, $\dfrac{f(aR)}{a} \to 0$

Along the path $\quad x_2 = x_1^2$, we see that

$$f(x_2, x_1) \to 1/2 \text{ as } x_1 \to 0$$

and along the path $\quad x_2 = x_1$,

$$f(x_1, x_2) \to 0 \text{ as } x_1 \to 0.$$

**Example 1.1.3**  We now consider $f: R^2 \to R$ defined by

$$f(x, y) = \begin{cases} \dfrac{xy}{x^2 + y^2} & \text{if } x \neq 0 \\[2mm] 0 & \text{if } x = 0 \end{cases}$$

$$(Df)\,(0)\,h = \lim_{a \to 0} \frac{f(0 + ah) - f(0)}{a}$$

$$= \underset{a \to 0}{\text{Lt}}\ \left. \frac{a^2 h_1 h_2}{a^2 h_1^2 + a^2 h_2^2} \right| a$$

$$= \text{Lt}\ \frac{h_1 h_2}{a\left(h_1^2 + h_2^2\right)}$$

This does not exist unless $h = (0, h_2)$ or $h = (h_1, 0)$.

So Gateaux derivative of the above function does not exist in any other direction, it exists in the direction of $x$ and $y$ axis.

**Example 1.1.4**  $\quad f(x) = \begin{cases} \dfrac{x_1^3\, x_2}{x_1^4 + x_2^2} & \text{if } x \neq (0, 0) \\[2mm] 0 & \text{if } x = (0, 0) \end{cases}$

$f$ is Gateaux differentiable at 0 with Gateaux derivative 0. But it is not Frechet differentiable.

For,

$$\frac{\|f(h)\|}{\|h\|} = \frac{\left|h_1^3 h_2\right|}{\left(h_1^4 + h_2^2\right)} \left[\frac{1}{\sqrt{h_1^2 + h_2^2}}\right.$$

$$= \frac{1}{\sqrt{1 + h_1^2}}\ \text{for } h_2 = h_1^2$$

So RHS $\to 1$ if $h_1 \to 0$.

$$\underset{a \to 0}{\text{Lt}}\ \frac{f(ah)}{a} = \underset{a \to 0}{\text{Lt}}\ \frac{a h_1^3 h_2}{(a^2 h_1^4 + h_2^2)} \to 0.$$

**Theorem 1.1.3**  If $f$ is Frechet differentiable at a point $a \in A$ then $\exists$ a unique linear map of $X$ into $Y$ which satisfies the condition.

**Proof.**  Let $T, S$ both satisfy the condition and let $\epsilon > 0$. $\exists\ \delta_1 > 0,\ \delta_2 > 0$ such that

$$\|f(x) - f(a) - T(x - a)\| \leq \epsilon\ \|x - a\|$$

for $\qquad \|x - a\| < \delta_1$ and

$$\|f(x) - f(a) - S(x - a)\| \le \in \|x - a\|$$

for $\qquad \|x - a\| < \delta_2.$

Let $\delta = \min (\delta_1, \delta_2)$ and $\|x - a\| < \delta$.

Then

$$\|(T - S)(x - a)\| = \|T(x - a) - S(x - a)\|$$
$$\le \|T(x - a) - f(x) + f(a)\|$$

If $y \in X$ and $y \ne 0$ then

$$\left\|\left(\frac{\delta y}{2\|y\|} + a\right) - a\right\| < \delta \text{ and so}$$

$$\left\|(T - S)\left(\frac{\delta y}{2\|y\|}\right)\right\| \le 2\in \left\|\frac{\delta}{2\|y\|} y\right\| = 2\delta\in.$$

This gives $\|(T - S)y\| \le 2\in\|y\|$. The last inequality is clearly true when $y = 0$ and hence $\|T - S\| \le 2 \in$ and $T - S \in B(X, Y)$. Since $\in$ is arbitrary, $\|T - S\| = 0$ and hence $T = S$.

This completes the proof.

To simplify the notation we shall write $f'(a) . x$ instead of $(f'(a))x$.

**Theorem 1.1.4**   Let $f$ be differentiable at $a \in A$. Then $f'(a) \in B(X, Y)$ iff $f$ is continuous at $a$.

**Corollary 1.1.1**   Let $X$ be finite dimensional. Let $f$ be differentiable at $a \in A$. Then $f$ is continuous at $a \in A$.

**Proof of Corollary.** $f'(a)$ is a linear map. $X$ is finite dimensional. Hence every linear map is bounded; so $f'(a)$ is bounded. Therefore $f$ is continuous at $a \in A$.

**Proof of the Theorem 1.1.4**   Let $f$ be continuous at $a \in A$ and let $\in > 0$. $\exists \delta_1 > \delta_2 > 0$ such that

$$\|f(x) - f(a)\| < \frac{\in}{2} \text{ for } \|x - a\| < \delta_1$$

and

$$\|f(x) - f(a) - f'(a) . (x - a)\| \le \frac{\in}{2}\|x - a\|$$

for $\qquad \|x - a\| < \delta_2.$ Let $\delta = \min (\delta_1, \delta_2, 1),$

and $\qquad \|x - a\| < \delta$. Then

$$\| f'(a).(x - a)\| \leq \|f'(a).(x - a) - f(x) + f(a)\| + \|f(x) - f(a)\|$$

$$< \frac{\epsilon}{2} \, \|x - a\| + \frac{\epsilon}{2} < \epsilon.$$

This show that $f'(a)$ is continuous at $a$. Hence $f'(a)$ is bounded at $a$.
Conversely, suppose that $f'(a) \in B\ (X,\ Y)$. Then $\exists \delta > 0$ such that

$$\|f(x) - f(a) - f'(a)(x - a)\| \leq \|x - a\| \text{ for } \|x - a\| < \delta.$$

Let $\|x - a\| < \delta$. Then

$$\|f(x) - f(a)\| \leq \|f(x) - f(a).(x - a)\| + \|f'(a).(x - a)\|$$

$$\leq \|x - a\| + \|f'(a)\| \, \|x - a\|$$

$$= (1 + \|f'(a)\|) \, \|x - a\|$$

Thus $f$ is continuous at $a$.

### 1.1.3   Partial Derivatives

**Theorem 1.1.5**   If $X_1$ and $X_2$ are real nlss, then $X_1 \ X \ X_2$ is a real nls with

$$(x_1,\ x_2) + (y_1,\ y_2) = (x_1 + y_1,\ x_2 + y_2)$$
$$a\,(x_1,\ x_2) = (ax_1,\ ax_2)$$
$$\|(x_1,\ x_2)\| = \max\ (\|x\|,\ \|x_2\|)$$

Let $A$ open $\subset X_1 \ X \ X_2$ and $f : A \to Y$.
Let $(a_1,\ a_2) \in A$,

$$A_1 = \{x_1 \in X_1 : (X_1,\ a_2) \in A\}$$
$$g\,(x_1) = f(x_1,\ a_2),\ \forall x_1 \in A_1.$$

Clearly $A_1$ is open in $X_1$.

**Definition 1.1.4**   $f$ is said to be differentiable w.r.t. the first variable at $(a_1,\ a_2)$ iff $g$ is differentiable at $a_1$. If $f$ is differentiable at $a_1$ we write $g'(a_1) = f'_1(a_1,\ a_2)$. The derivative $f'(a_1,\ a_2)$ is called the partial derivative of $f$ w.r.t. the first variable at $(a_1,\ a_2)$, it is a linear map of $X_1$ into $Y$.

**Theorem 1.1.6**   If the Frechet differential of $F$ exists at $x$, then the Gateaux differential exists at $x$ and they are equal.

**Proof.**   Let the Frechet differential of $F$ be $\delta F(x;\ h)$. By definition we have for any $h$,

$$\frac{1}{a} \, \|F(x + ah) - F(x) \ \delta F(x;\ ah)\| \to 0$$

as $\alpha \to 0$. Thus by the linearity of $\delta F(x; \alpha h)$ with respect to $\alpha$,

$$\lim_{a \to 0} \frac{F(x + ah) - F(x)}{a} = \delta F(x; ah) \to 0$$

and this completes the proof.

**Theorem 1.1.7**   If the Gateaux derivative $F'(x)$ of $F$ exists in some neighbourhood of the point $x$ and is continuous at $x$, then $F$ is also Frechet differentiable at $x$ and both are equal.

**Proof.**   Put $w(x, h) = F(x + h) - F(x) - F'(x)h$.

Then

$$(e, w(x, h)) = (e, F(x + h) - F(x))$$

$$- (e, F'(x)h), \ e \in Y^*$$

By Mean–Value Theorem we get

$$(e, w(x, h)) = (e, F'(x + \theta h)h - F'(x)h), \ 0 < \theta < 1$$

By Hahn Banach theorems $e$ can be so chosen that

$$\|w(x, h)\| = |e, w(x, h))| \text{ and } \|e\| = 1$$

Hence we get

$$\|w(x, h)\| \leq \| F'(x + \theta h) - F'(x)\| \ \|h\|$$

This implies that

$$\frac{\|w(x, h)\|}{\|h\|} \leq \|F'(x + \theta h) - F'(x)\|$$

Since $F'(x)$ is continuous, the RHS of the above inequality $\to 0$ as $\|h\| \to 0$ and this proves our theorem.

**Theorem 1.1.8**   Let $X, Y, Z$ be Banach spaces. Let $G:X \to Y$. Gateaux differentiable at $x$ and $F:Y \to Z$ Frechet differentiable at $G(x)$.

Then $\qquad\qquad H = F.G.$ is Gateaux differentiable at $x$ and.

$$H'(x) = (F.G)'(x) = F'(G(x)) \ G'(x).$$

If $G$ is also Frechet differentiable, then $H$ is Frechet differentiable.

**Proof.**   For $t \neq 0$ we have

$$\frac{1}{|t|} \|H(x + th) - Hx - tF'(Gx) \ G'(x)h\|$$

$$\leq \frac{1}{|t|} \|F(G(x + th)) - F(G(x))$$

$$- tF'(Gx)\,(G(x + th) - G(x) - tG'(x)\,h\| \text{-----}(*)$$

Since $G$ is Gateaux differentiable, second term of RHS of $*$ $\to 0$ as $t \to 0$.

For any value of $t$ such that $G(x + th) \neq G(x)$, the first term may be multiplied and divided by $\|G(x + th) - G(x)\|$. Now since $F$ is Frechet differentiable and $\|G(x + th) - G(x)\| \to 0$, the first term on RHS of $*$ $\to 0$. Hence LHS of $*$ $\to 0$.

This implies that $H$ is Gateaux differentiable.

In a similar manner, it can be shown that $H$ is Frechet differentiable if $G$ is so.

**Definition 1.1.5**  Let $f$ be a real functional defined on a subset $K$ of a nls $X$. A point $x_0 \in K$ is said to be a (relative) minimum of $f$ on $K$ if $\exists$ an open sphere $N$ containing $x_0$ such that $f(x_0) \neq f(x)\ \forall x \in K \cap N$, and a strict (relative) minimum if strict inequality holds for $x \neq x_0$, $x \in K \cap N$.

We define (relative) maximum in a similar way.

A point, which is either a minimum or a maximum is called an extremum.

A relative extremum is also called a local extremum.

**Theorem 1.1.9**  Let $f$ be a real Gateaux differentiable functional defined on a real vector space $X$. If $x_0 \in X$ is an extremum for $f$, then $\delta f(x_0; h) = 0\,\forall h \in X$.

**Proof.**  Let $f$ be Gateaux differentiable let $x_0$ be an extremum for $f$.

Define for every $h \in X$, a function $F : R \to R$ by

$$F(a) = f(x_0 + ah).$$

Then $F$ must have an extremum at $\alpha = 0$. Therefore by ordinary calculus,

$$\frac{d}{da} F(a)\bigg|_{a=0} = \frac{d}{da} f(x_0 + ah)\bigg|_{a=0} = 0$$

This completes the proof.

## 1.2  INTEGRATION

Let $X$ be a Banach space and $X^*$ be its dual. Consider a function $\phi : [0, 1] \to X$. Now we shall give the meaning to the expression

$$\int_0^1 \varphi(t)\, dt$$

**Definition 1.2.1**  The above integral if exists, is the unique vector $x \in X$ such that

$$f(x) = \int_0^1 f(\varphi(t))\,dt \qquad\qquad \dots(1.1)$$

for every $f \in X^*$.

**Remark 1.2.1**   The vector $x \in X$ satisfying (1.1) is unique, for if there exist two vectors $x_1$ and $x_2$, then $f(x_1) = f(x_2)$ for all $f \in X^*$, which implies $x_1 = x_2$ by Hahn–Banach theorem. We note that the integrand in (1.1) is a real valued function and hence the integral can be defined in the usual way.

Our next theorem shows that if $\phi$ is continuous then the integral exists.

**Theorem 1.2.1**   Let $\phi : [0, 1] \to X$ be continuous. Then

$$\int_0^1 \varphi(t)\,dt \text{ exists. Further,}$$

$$\left\| \int_0^1 \varphi(t)\,dt \right\| \le \int_0^1 \|\varphi(t)\|\,dt.$$

**Proof.**   Let $n \ge 1$ be an integer and set

$$x_n = \frac{1}{2^n} \sum_{i=0}^{2^n-1} \varphi\left(\frac{i}{2^n}\right)$$

We claim that $\{x_n\}$ is a Cauchy sequence in $X$. Let $\epsilon > 0$. Since $\phi$ is uniformly continuous, there exists a $\delta > 0$ such that

$$|t - s| < \delta \implies \|\varphi(t) - \varphi(s)\| < \epsilon.$$

Choose $n_0$ so that $\dfrac{1}{2^n} < \delta$ for all $n \ge n_0$.

$$y_n - y_{n+m} = \frac{1}{2^n} \sum_{0}^{2^n-1} \varphi\left(\frac{i}{2n}\right) - \frac{1}{2^{n+m}} \sum_{0}^{2^{n+m-1}} \varphi\left(\frac{i}{2^{n+m}}\right)$$

$$= \frac{1}{2^n} \sum_{i=0}^{2^n-1} \left[ \varphi\left(\frac{i}{2^n}\right) - \frac{1}{2^m} \sum_{j=0}^{2^m-1} \varphi\left(\frac{i2^m + j}{2^{n+m}}\right) \right]$$

$$= \frac{1}{2^{n+m}} \left[ \sum_{i=0}^{2^n-1} \sum_{j=0}^{2^m-1} \varphi\left(\frac{i}{2^n}\right) + \left(\frac{i2^m + j}{2^{n+m}}\right) \right]$$

Since

$$\left| \frac{i}{2^n} - \frac{i2^m + j}{2^{n+m}} \right| \le \frac{2^{m-1}}{2^{n+m}} < \delta,$$

We have

$$\left\| \varphi\left(\frac{i}{2^n}\right) - \varphi\left(\frac{i2^m + j}{2^{n+m}}\right) \right\| < \epsilon$$

and hence $||x_n - x_{n+m}|| < \epsilon$ for all $m \geq 0$. Thus $\{x_n\}$ is Cauchy and hence has a limit, say $x \in X$. Now if $f \in X^*$,

$$f(x) = \lim_{n \to a} f(x_n)$$

$$= \lim_{n \to a} \frac{1}{2^n} \sum_{i=0}^{2^n-1} f\left(\varphi\left(\frac{i}{2^n}\right)\right)$$

$$= \int_0^1 f(\varphi(t))\, dt$$

by the usual definition of the Riemann integral Hence the integral $\int_0^1 \varphi(t)\,dt$ exists.

For this unique $x$ there exists, by Hahn Banach theorem, $f \in X^*$ such that $||f|| \leq 1$ and $f(x) = ||x||$.

That is, $$\left\| \int_0^1 \varphi(t)\,dt \right\| = \int_0^1 f(\varphi(t))\,dt \leq \int_0^1 |f(\varphi(t))|\,dt \leq \int_0^1 ||\varphi(t)||\,dt$$

The proof is complete.

**Remark: 1.2.2**   Similarly one can define the integral

$$\int_a^b \varphi(t)\,dt$$

for any continuous function $\varphi:[a, b] \to X$. If $\varphi: [a, \infty] \to X$, then we define.

$$\int_0^\infty \varphi(t)\,dt = \lim_{M \to \infty} \int_0^M \varphi(t)\,dt$$

Similarly we have the following:

(i) $\int_{-\infty}^\infty \varphi(t)\,dt = \int_{-\infty}^a \varphi(t)\,dt,\ a \in R$

(ii) If $a < b < c,\ a, b, c \in R \cup \{-\infty, \infty\}$ and if the relevant integrals exist,

$$\int_a^c \varphi(t)\,dt = \int_a^b \varphi(t)\,dt + \int_b^c \varphi(t)\,dt$$

Our next theorem generalizes the relation (1.2.1).

**Theorem 1.2.2**   Let $\varphi:[a, b] \to X$ be continuous and $T : X \to Y$ be a bounded linear operator where $Y$ is also a Banach space. Then

$$T\left(\int_0^1 \varphi(t)dt\right) = \int_0^1 T(\varphi(t)dt) \qquad\qquad ...(1.2)$$

**Proof.**    Let $f \in Y^*$. Then $f \circ T \in Y^*$.

$$\text{Then } f \circ T\left(\int_a^b \varphi(t)\,dt\right) = \int_0^1 (f \circ T)\,(\varphi(t)dt)$$

$$(\text{i.e.}) \qquad f\left|T\int_a^b \varphi(t)dt\right| = \int_0^1 f(T(\varphi(t)))dt$$

This implies the relation (1.2).

# Convex Sets

Let $X$ be a real vector space. For $x, y \in X$, the line through $x, y$ is defined as the set.

$$\{\lambda x + (1 - \lambda) y : \lambda \in R\}$$

or equivalently,

$$\{\lambda x + \mu y : \lambda, \mu \in R, \lambda + \mu = 1\}.$$

We can rewrite this as

$$\{y + \lambda(x - y) : \lambda \in R\}$$

If $X = R^2$, then it is obvious that the above represents the line through $x$ and $y$.

For $x, y \in X$ we define the line segment joining $x, y$ as follows:

The closed line segment

$$= \{\lambda x + (1 - \lambda)y : 0 \leq \lambda \leq 1\}$$

The open line segment

$$= \{\lambda x + (1 - \lambda)y : 0 < \lambda < 1\}.$$

**Definition 2.1.1**  *A subset $K$ of $X$ is said to be convex if $x_1, x_2 \in K$ and $0 \leq \lambda \leq 1$ imply that $\lambda x_1 + (1 - \lambda)x_2 \in K$. In other words, the line segment joining $x_1$ and $x_2$ lies entirely in $K$.*

**Definition 2.1.2**  *A set $K \subset X$ is called star-like (or star-shaped) at a point $x_0 \in X$ if the line segment joining $x_0$ and any point $x \in K$ lies entirely in $K$, i.e., if*

$$\text{for } x \in K, 0 \leq \lambda \leq 1,$$

$$\lambda x + (1 - \lambda) x_0 \in K$$

*Remark 2.1.1*  Note that $K$ is convex iff it is star-like at each of its points.

## 2.1  SOME EXAMPLES OF CONVEX SETS

1. The null set $\phi$ is convex
2. Every singleton set is convex
3. The whole vector space is convex
4. Every subspace of $X$ is convex
5. Let $X$ be a real nls.

   Then every open ball

$$B_r(x) = \{y \in X : \|y - x\| < r\}$$

is a convex set.

Every closed ball

$$B_r[x] = \{y \in X : \|y - x\| \le r\}$$

is a convex set as well.

A sphere

$$S_r(x) = \{y \in X : \|y - x\| = r\}$$

is not a convex set.

6. The intersection of a family of convex sets in a real vector space is convex.

7. If $A$ and $B$ are any two convex sets in $X$, then the set

$$A + B = \{z : z = x + y,\ x \in A,\ y \in B\}$$

is convex.

8. If $\lambda \in R$ and $A$ is a convex set in $X$, then

$$\lambda A = \{z : z = \lambda x,\ x \in A\}$$

is also a convex set.

**Proof.**

6. Let $\{K_i : i \in I\}$ be a family of convex sets in $X$ and let $K = \bigcap_{i \in I} K_i$. Let $x, y \in K$ and $0 \le \lambda \le 1$. Then $x, y \in K_i$ for each $i \in I$. Since $K_i$ is convex, it follows that $\lambda x + (1 - \lambda) y \in K_i$ for each $i$. Hence $\lambda x + (1 - \lambda) y \in K$ and this completes the proof.

9. In $R^n$ we define a plane as follows:

   For $c \in R^n$, $c \ne 0$ and $\alpha \in R$, the set

$$\{x \in R^n : cx = \alpha\}$$

is called a plane.

   Each plane in $R^n$ is a convex set.

10. Half-space in $R^n$

Let $c \in R^n$, $c \neq 0$ and $\alpha \in R$.

The set

$$\{x \in R^n : cx < \alpha\}$$

is called an open half space and the set

$$\{x \in R^n : cx \leq a\}$$

is called a closed half space. Each half space is a convex set in $R^n$. In $R^n$, let

$$e_0 = (0,\ 0,\ 0,\ ....\ 0)$$
$$n \text{ zeros}$$
$$e_1 = (0,\ 1,\ 0,\ 0,\ ...,\ 0)$$
$$e_2 = (0,\ 0,\ 1,\ 0,\ 0,\ ...,\ 0)$$

$$\cdot$$
$$\cdot$$
$$\cdot \qquad\qquad n\text{th plane}$$
$$e_n = (0,\ 0,\ ...,\ 1,\ 0,....,\ 0)$$

Let $\Delta_n$ be the convex set whose vertices are $e_0, e_1, ..., e_n$.
Each $\Delta_n$ is called a simplex of dimension $n$ or an $n$–simplex.
Note that

$\Delta_0$ is the single point $e_0$,
$\Delta_1 = $ the line segment joining $e_0$ and $e_1$,
$\Delta_2 = $ the triangle with vertices $e_0, e_1, e_2$,
$\Delta_3 = $ the tetrahedron,

$$\cdot$$
$$\cdot$$
$$\cdot$$

$\Delta_n = $ the $n$-dimensional polyhedron

**Definition 2.1.3** Let $A$ be a subset of a nls $X$. We have already seen that the intersection of all convex subsets of $X$ containing $A$ is convex. It is, in fact, the smallest convex subset of $X$ containing $A$.

This set is called the convex hull of $A$ and is denoted by $c_0 A$. To be precise

$$c_0 A = h\ \{A_i : A_i \text{ is convex and } A \subseteq A_i\}.$$

Let $X$ be a real vector space.

**Definition 2.1.4** A linear variety $V$ in $X$ is a translation of a subspace $M$, i.e., if $x_0 \in X$ and $x_0 \notin M$, then a linear variety

$$V = x_0 + M.$$

**Definition 2.1.5** *A hyperplane $H$ in $X$ is a maximal proper linear variety, i.e., a linear variety $H$ such that $H \neq X$ and if $V \supset H$ is a linear variety, then either $V = H$ or $V = X$.*

**Examples 2.1.1**  (i) Lines are hyperplanes in $R^2$,

(ii) Planes are hyperplanes in $R^3$.

**Theorem 2.1.1**  Let $H$ be a hyperplane in a linear space $X$. Then $\exists$ a linear functional $f$ on $X$ and a real constant $c$, such that $H = \{x : f(x) = c\}$. Conversely, if $f$ is a nonzero linear functional on $X$, then the set $\{x : f(x) = c\}$ is a hyperplane in $X$.

**Proof.**  Let $H$ be a hyperplane in $X$. Then $H$ is the translation of a subspace $M$ in $X$, say $H = x_0 + M$. There are two possible cases: $x_0 \notin M$ and $x_0 \in M$.

***Case I:***  Let $x_0 \notin M$, then $[M + x_0] = X$. For $x = ax_0 + m$, with $m \in M$ define

$$f(x) = \alpha.$$

Then $\qquad\qquad H = \{x : f(x) = 1\}.$

***Case II:***  Let $x_0 \in M$. Then take $x_1 \notin M$, $[M + x_1]$, $H = M$. Define for

$$x = \alpha x_1 + m,$$

$$f(x) = \alpha.$$

Then $\qquad\qquad H = \{x : f(x) = 0\}.$

Conversely, let $f$ be a nonzero linear functional on $X$ and let

$$M = \{x : f(x) = 0\}$$

Clearly $M$ is a subspace. Let $x_0 \in X$ with $f(x_0) = 1$. Then for any $x \in X$,

$$f[x - f(x) x_0] = f(x) - f(x) f(x_0)$$

$$= f(x) - f(x) = 0,$$

and hence $x - f(x) x_0 \in M$.

Thus $X = [x_0 + M]$, and $M$ is a maximal proper subspace. For any $c \in R$, let $x_1$ be any element for which $f(x_1) = c$. Then

$$\{x : f(x) = c\} = \{x : f(x - x_1) = 0\},$$

$$M + x_1,$$

which is hyperplane and this completes the proof.

*Remark 2.1.2*  Observe that a hyperplane $H$ in a nls $X$ is either closed or dense. Because, since $H$ is a maximal linear variety, either $\overline{H} = H$ or $\overline{H} = X$. Hence either $H$ is closed or $H$ is dense.

We are primarily interested in closed hyperplanes and these hyperplanes correspond to the bounded (continuous) linear functionals on $X$. We have

**Theorem 2.1.2**  Let f be a nonzero functional on a nls $X$. Then the hyperplane $H = \{x : f(x) = c\}$ is closed for every $c$ iff $f$ is continuous.

**Proof.**  Let $f$ be continuous. Let $\{x_n\}$ be a sequence in $H$ which converges to an element $x \in X$. Then $c = f(x_n) \rightarrow f(x)$; thus $x \in H$ and hence $H$ is closed.

Let $X = [x_0 + M]$ and let $x_n \rightarrow x$ in $X$. Then

$$x_n = \alpha_n x_0 + m_n, \quad x = \alpha x_0 + m, \quad \text{let } d \text{ denote the distance of } x_0 \text{ from } M.$$

Since $M$ is closed, $d > 0$. We now have

$$|\alpha_n - \alpha|\, d \leq \|x_n - x\| \rightarrow 0$$

and hence $\alpha_n \rightarrow \alpha$.

Also
$$\begin{aligned}
f(x_n) &= f(\alpha_n x_0 + m_n) \\
&= \alpha_n\, f(x_0) + f(m_n) \\
&= \alpha_n\, f(x_0) \\
\rightarrow \alpha f(x_0) &= f(x).
\end{aligned}$$

Thus $f$ is continuous and this completes the proof.

Therefore linear functionals are visualized as hyperplanes generated in the primal space rather than the elements of the dual space.

This leads to the following definition:

**Definition 2.1.6**  If

$$H = \{x : f(x) = c\}$$

is a hyperplane in $X$, then the sets

$$H_1 = \{x \in X : f(x) \leq c\}$$

and
$$H_2 = \{x \in X : f(x) \geq c\}$$

are called half spaces determined by $H$.

**Definition 2.1.7**  A point $x_0$ of a convex set $K$ is called an extreme point if there do not exist points $x_1, x_2 \in K$ and $\alpha$ with $0 < \alpha < 1$ such that

$$x_0 = \alpha x_1 + (1 - \alpha)x_2.$$

**Examples 2.1.2**  The extreme points of a closed ball are its boundary points. A half space has no extreme point.

**Definition 2.1.8**   Let $X$ be a vector space and let $A$ and $B$ be subsets of $X$. $A$ and $B$ are said to be separated by the hyperplane $H = \{x : f(x) = \alpha\}$ if $A$ (resp. $B$) is in the half space $\{x : f(x) \le \alpha\}$ and $B$ (resp. $A$) is in the half space $\{x : f(x) \ge \alpha\}$.

A and $B$ are said to be strictly separated or strongly separated by the hyperplane $H$ if $A$ (resp. $B$) is in $\{x : f(x) < \alpha\}$ and $B$ (resp. $A$) is in $\{x : f(x) > \alpha\}$.

For Example, the tangent line to a circle $R^2$ separates the closed disc from the point of tangency.

**Definition 2.1.9**   Let $K$ be a convex set in a nls $X$ and let $x$ be an interior point of $K$. Then the Minkowski functional $p$ of $K$ is defined on $X$ by

$$p(x) = \inf \left\{ r : \frac{x}{r} \in K, r > 0 \right\}$$

Note that if $K$ is the unit sphere, then

$$p(x) = \|x\|.$$

**Theorem 2.1.3**   **(Mazur: Geometric Form of Hahn-Banach Theorem)**

Let $X$ be a real nls. $K$ a convex set in $X$ with int $K \ne \phi$. $V$ a linear variety in $X$, $V \cap$ int $K = \phi$.

Then $\exists$ a closed hyperplane in $X$ containing $V$ and not containing any interior point of $K$, i.e., $\exists f \in X^*$ and a constant $c$ st.

$$f(v) = c \qquad\qquad \forall v \in V$$

$$f(k) < c \qquad\qquad \forall k \in \text{int } K.$$

**Proof.**   Without loss of generality, we may assume that $0 \in$ int $K$; [for, otherwise we can translate $K$ appropriately].

Let $M$ be a subspace of $X$ generated by $V$. Then $V$ is a hyperplane in $M$ and $0 \notin V$. Thus $\exists$ a linear functional $f$ on $M$ st

$$V = \{x \in X : f(x) = 1\}.$$

Let $p$ be the Minkowski's functional of $K$. Since $V \cap$ int $K = \Phi$, we have

$$f(x) = 1 \le p(x) \; \forall x \in V.$$

Now

$$f(\alpha x) = \alpha f(x) = \alpha \le p(\alpha x), \; \alpha > 0, \; x \in V$$

$$f(\alpha x) \le 0 \le p(\alpha x), \; \alpha < 0, \; x \in V$$

Thus

$$f(x) \le p(x) \; \forall x \in M.$$

Therefore by extension form of Hahn-Banach Theorem, $\exists$ a linear functional $F$ on $X$ with

$$F(x) \leq p\,(x) \ \forall x \in X.$$

Define

$$H = \{x : F(x) = 1\}$$

Since $F(x) \leq p(x)$ since $p$ is continuous, $F$ is continuous, $F(x) < 1 \ \forall x \in$ int $K$.

Therefore $H$ is the desired closed hyperplane and this completes the proof.

**Definition 2.1.10**  A closed hyperplane $H$ in a nls $X$ is said to be a support (or a supporting hyperplane) for the convex set $K$ if $K$ is contained in one of the closed half-spaces determined by $H$ and $H$ contains a point of $K$.

**Theorem 2.1.4**  (Support Theorem)

If $(i)$ $x \notin$ int $K \neq \Phi$, then $\exists$ a closed hyperplane $H \ni x$ st $K$ lies on one side of $H$.

**Theorem 2.1.5**  Let $K_1$ and $K_2$ be convex sets in $X$ st

$$\text{int } K_1 \neq \Phi, \ K_2 \cap \text{ int } K_1 = \Phi.$$

Then $\exists$ a closed hyperplane $H$ separating $K_1$ and $K_2$, i.e., $f \in X^*$ st

$$\sup_{x \in K_1} f(x) \leq \inf_{x \in K_2} f(x).$$

In other words, $K_1$ and $K_2$ lie in opposite half spaces determined by $H$.

**Proof.**  Let $K = K_1 - K_2$; then

$$\text{int } K \neq \Phi \text{ and } 0 \notin \text{ int } K.$$

So by above theorem $\exists f \in X^*, f \neq 0$, st $f(x) \neq 0 \ \forall x \in K$.

Thus for $x_1 \in K_1, x_2 \in K_2,$

$$f(x_1) \neq f(x_2)$$

Consequently $\exists c \in R$ st $\displaystyle \sup_{k_1 \in K_1} f(K_1) \leq c \leq \inf_{k_2 \in K_2} f(K_2).$

Thus the desired hyperplane is

$$H - \{x : f(x) = c\}.$$

**Theorem 2.1.6**  If $K$ is a closed convex set and $x \notin K$, $\exists$ a closed halfspace that contains $K$ but does not contain $x$.

**Proof.**  Let $d = \displaystyle\inf_{k \in K} \|x - k\|.$

Since $K$ is closed $d > 0$.

Let
$$S \equiv S_{d/2}(x).$$

Then applying above theorem to $S$ and $K$ we obtain the result.

This theorem can be restated as follows:

A closed convex set in a nls is equal to the intersection of all the closed half spaces that contain it.

## 2.2  TOPOLOGICAL PROPERTIES

In the following we shall discuss some topological properties of convex sets in $R^n$. Before proving the theorems we quote some definitions which will be needed in the sequel.

**Definition 2.2.1**  Let $X$ and $Y$ be any two topological spaces. Let $f$ and $g$ be maps (continuous maps) from $Y$ into $X$. Let $I$ denote the unit interval $[0, 1]$. We say that $f$ and $g$ are homotopic relative to a nonempty subset $A$ of $Y$ iff there exists a continuous function

$$F : YXI \rightarrow X$$

such that

$$F(y, 0) = f(y), \quad y \in Y$$
$$F(y, 1) = g(y), \quad y \in Y$$
$$F(a, t) = f(a) = g(a), a \in A, t \in I.$$

If $f$ is homotopic to $g$ relative to $A$, we write this as

$$f \cong g \ (\text{rel } A).$$

If $\qquad A = \Phi, \quad$ we simply have

$$f \cong g.$$

**Definition 2.2.2**  A (continuous) map $\sigma : I = [0, 1] \rightarrow X$ is called a path. Two paths $\sigma_1$ and $\sigma_2$ are called homotopic if $\exists$ a continuous function.

$$F : IXI \rightarrow X \text{ st}$$
$$F(s, 0) = \sigma_1 (s), \qquad F(s, 1) = \sigma_2 (s)$$
$$F(0, t) = x_0, \qquad F(1, t) = x_1.$$

**Theorem 2.2.1**  Homotopy of maps is an equivalence relation.

**Proof.**  (i) Define $F : Y X I \rightarrow X$ by

$$F(y, t) = f(y)$$

Thus $f \cong f.$ (reflexive)

(ii) Given $f \cong g$ i.e. $F : Y X I \rightarrow X$ st

$$F(y, 0) = f(y)$$

$$F(y, 1) = g(y)$$

$$F(a, t) = f(a) = g(a)$$

Define

$$G : Y \times I \to X \text{ by}$$

$$G(y, t) = F(y, 1-t)$$

Then

$$G(y, 0) = F(y, 1) = g(y)$$

$$G(y, 1) = F(y, 0) = f(y)$$

$$G(a, t) = F(a, 1-t) = g(a) = f(a)$$

Thus $g \cong f$.

(iii) Let $f \cong g$ (via $F$), $\quad g \cong h$ (via $G$).

Define $H : Y \times I \to X$ by

$$H(y, t) = \begin{cases} F(y, 2t), \ 0 \le t \le \dfrac{1}{2} \\ G(y, 2t - 1), \ \dfrac{1}{2} \le t \le 1 \end{cases}$$

Then $f \cong h$. (transitivity).

**Example 2.2.1**   Let $X = Y = R^n$.

Define $F : R^n \times I \to R^n$ by

$$F(x, t) = tx.$$

Then $F(x, 0) = 0 = 0(x)$, where $0$ denotes the zero map.

$$F(x, 1) = x = id(x)$$

$$F(0, t) = 0 = 0(o) = id(o).$$

Thus $\qquad\qquad 0 \cong id$.

**Definition 2.2.3**   Let $X$ be a topological space and let $x_0 \in X$.
The map

$$e_{x0} : X \to X \text{ defined by}$$

$$e_{x0}(x) = x_0 \in X \text{ for all } x \in X,$$

is called a constant map (the constant map with value $X_0 \in X$).

**Definition 2.2.4**   A topological space is contractible if the identity map *id*: $X \to X$ is homotopic to some constant map $ex_0$: $X \to X$.

**Theorem 2.2.2**  Every convex set $K \subseteq R^n$ ($n \geq 1$) is contractible.

**Proof.**  Define a map

$$F : K X I \to K \text{ by}$$

$$F(x, t) = tx + (1-t) x_0, \quad \forall x \in K$$

where $x_0 \in K$ is some fixed element.

Since $K$ is a convex set and $x_0 \in K$, $tx + (1-t) x_0 \in K$ for all $x \in K$ and hence $F$ is well-defined.

Now $\qquad F(x, 0) = x_0 = ex_0(x), \quad \forall x \in K$

where $ex_0$ denotes the constant map with value $x_0$ and

$$F(x, 1) = x = id(x)$$

where $id$ denotes the identity map.

Thus $\qquad ex_0 \cong id$

and this proves the theorem.

**Remark 2.2.1**  If a set $K \subseteq R^n$ is star-like at some point $x_0 \in K$, then it is contractible.

**Definition 2.2.5**  A topological space $X$ is said to be path-connected if any pair of points is connected by a path; in otherwords, if for any pair of points $x_0, x_1, \exists$ *a* continuous map $\alpha : I \to X$ st $\alpha(0) = x_0$, $\alpha(1) = x_1$.

**Definition 2.2.6**  A topological space is said to be disconnected if there exists two open sets $A$ and $B$ such that

$$X = A \cup B.$$

Otherwise $X$ is said to be connected.

It is known that

(i) every contractible space is path-connected,

(ii) every path-connected space is connected.

Therefore it follows that every convex set in $R^n$ is path-connected and hence connected.

We can also independently prove in the following manner that every convex set in $R^n$ is connected.

**Theorem 2.2.3**  Every convex set in $R^n$ is connected.

**Proof.**  Let $K$ be a convex set in $R^n$. Let $x_0 \in K$ be fixed.

Consider the set

$$Ax = \{ tx + (1-t)x_0 : t \in I = [0, 1] \}$$

This set $A_x$ is connected.

$$K = \bigcup_{x \in K} A_x \text{ and } x_0 \in \bigcup_{x \in K} A_x.$$

Since the union of a family of connected sets with nonempty intersection is connected, $K$ is connected.

## 2.3 CONVEX CONES

Let $X$ be a normed linear space $X^*$ be its dual, that is the space of all continuous linear functionals on $X$. It is well known that $X^*$ is a Banach space under the norm defined by

$$\|x^*\| = \sup_{\|x\| \le 1} |x^*(x)|$$

There is natural duality between $X$ and $X^*$ determined by the bilinear functional

$$(.,\,.) : X \, X \, X^* \to R$$

$$(x, x^*) = x^*(x) \quad \forall x \in X, x^* \in X^*$$

**Definition 2.3.1** A set $K \subseteq X$ is said to be a cone if $x \in K$ and $\lambda \ge 0$ $\Rightarrow \lambda x \in K$. $K$ is said to be a convex cone if it is a convex set as well as a cone.

**Definition 2.3.2** Let $S$ be a subset of a nls $X$. We define two subsets in $X^*$ as follows:

$$S^+ = \{x^* \in X^* : (x, x^*) \ge 0 \text{ for all } x \in S\}$$

and

$$S^- = \{x^* \in X^* : (x, x^*) \le 0 \ \forall x \in s\}$$

$S^+$ is called the positive conjugate (dual) cone of $S$ and $S^-$ is called the negative conjugate (dual) cone of $S$.

### Remark 2.3.1

(1) $S^+$ and $S^-$ are nonempty, since they always contain the zero functional.

(2) $S^+$ and $S^-$ are convex cones.

(3) If $S$ is a subspace, then

$$S^+ = S^- = S^1.$$

**Theorem 2.3.1** Let $S$ and $T$ be subsets of a nls $X$. Then

1. $S^+$ is a closed convex cone in $X^*$.

2. $S \subset T \Rightarrow S^+ \supset T^+$.

**Definition 2.3.3** Let $P$ be a convex cone in a vector space $X$. Then $P$ defines a partial order relation (por) in $X$ as follows:

We say that $x \geq y$ iff $x - y \in P$.

This is in fact a por can be checked easily, for,

(i) $x \geq x$   since $x - x = 0 \in P$.

(ii) $x \geq y$,   $y \geq x \Rightarrow x - y \in P$

and $y - x = -(x - y) \in P$.

But this is impossible unless

$$x - y = \ y - x = 0 \Rightarrow x = y.$$

(iii) $x \geq y$, $y \geq z \Rightarrow x - y \in P$, $y - z \in P$

$$\Rightarrow (x - y) + (y - z) = x - z \in P$$

$$\Rightarrow x \geq z.$$

Such a cone $P$ is called a positive cone.

The cone defined by

$$N = -P = \{ -x : x \in P\}$$

is called the negative come in $X$ with respect to $P$.

**Example 2.3.1**   $X = R^n$

$$P = R_+^n = \{x \in R^n : x = (x_1, x_2, \ldots, x_n):$$

$$x_i \geq 0, \ i = 1, 2, \ldots, n\}$$

This is called the non-negative orthant of $R^n$

**Definition 2.3.4**   Let $X$ be a nls $P \subset X$ be a positive convex cone. Then the polar or dual or conjugate cone of $P$ is defined by

$$p^+ = \{x^* \in X^* : (x, x^*) \geq 0 \ \forall x \in P\}$$

**Example 2.3.2**   If $P$ is $R_+^n$, then $P^+$ is the set of all linear functionals represented by elements of $R^n$ with non-negative components.

# Convex Functions

3

We start this chapter by introducing the definition of convex functional.

**Definition 3.1.1** Let $X$ be a real vector space and let $f$ be a mapping from $X$ into $R$. We define

$$epif = \{(x, \alpha) \in X \times R: f(x) \leq \alpha\}$$

$f$ is said to be convex if epif is a convex subset of $X \times R$.

The set epif is called the epigraph of $f$.

**Definition 3.1.2** f is said to be convex if for all $x, y \in X$ and for all $\lambda \in [0, 1]$,

$$f(\lambda x + (1 - \lambda) y) \leq \lambda f(x) + (1 - \lambda) f(y).$$

The equivalence of above two definitions will be established in Theorem 3.1.1. $f$ is said to be strictly convex if strict inequality holds in Definition 3.1.2 whenever $x \neq y$.

A functional $g$ defined on $X$ or on a convex subset of $X$ is said to be concave (strictly concave) if $^-g$ is convex (strictly convex).

## 3.1 EXAMPLES OF CONVEX FUNCTIONALS

1. $f : R \rightarrow R, f(x) = 3x + 4$
2. $f : R \rightarrow R, f(x) = |x|$
3. $f : R \rightarrow R, f(x) - x^2$
4. $f : [-\pi, 0] \rightarrow R, \qquad f(x) = \sin x$
5. $f : R \rightarrow R, \qquad f(x) = e^x$
6. $f : R^+ \rightarrow R, \qquad f(x) - \log x$
7. $f : R^+ \rightarrow R, \qquad f(x) = x \log x$
8. $f : R \rightarrow R, \qquad f(x) = x^2 - 2x$
9. $f : R^+ \rightarrow R, \qquad f(x) = - x^{1/2}$
10. $f : R^2 \rightarrow R,$

$$f(x_1, x_2) = 2x_1^2 + x_2^2 - 2x_1 x_2$$

11. $f : R^2 \to R$,

$$f(x_1, x_2) = x_1^2 + 3x_1 x_2 - 5y_1^2$$

12. $f : R^n \to R$

$$f(x_1, x_2, \ldots, x_n) = x_i, \qquad i = 1, 2, \ldots, n$$

13. The discontinuous function

$$f : [0, \infty) \to R$$

$$f(x) = \begin{cases} 1, & x = 0 \\ x^2, & x > 0 \end{cases}$$

14. Let $x$ be a nls. $f : X \to R$

$$f(x) = \|x\|.$$

15. Let $K$ be a convex subset in a Hilbert space $H$. Define the distance function

$$D(x, K) = \inf \{\|x - y\| : y \in K\}.$$

$D$ is a convex function.

16. Let $K$ be a convex set in a nls.

The gauge functional

$$X(x|K) = \inf \{\lambda \geq 0 : x \in \lambda K\}$$

17. The indicator function $I_k$

$$I_k(x) = \begin{cases} 0, & x \in K \\ \infty, & x \notin K. \end{cases}$$

**Example 3.1.1**   Let $A$ be a self-adjoint continuous operator on a Hilbert space $X$. Let $\phi$ be the quadratic functional defined by

$$\phi(x) = 1/2 \, (Ax, x)$$

$\phi$ is differentiable and $\nabla \phi = A$. $\phi$ is convex iff $A$ is positive semidefinite.

**Theorem 3.1.1**   f is convex iff epif is a convex subset of $X \times R$.

**Proof.**   $\Rightarrow$ Let $f$ be convex and let $(x, \alpha)$, $(y, \beta) \in$ epif.

Let $0 \leq \lambda \leq 1$. Then

$$f(\lambda x + (1 - \lambda) y) \leq \lambda f(x) + (1 - \lambda) f(y)$$

$$\leq \lambda \alpha + (1 - \lambda) \beta$$

Therefore

$$(\lambda x + (1 - \lambda)\, y,\ \lambda\alpha + (1 - \lambda)\, \beta)$$

$$= \lambda\, (x, \alpha) + (1 - \lambda)\, (y, \beta) \in \text{epif}$$

and hence epif is convex.

**Theorem 3.1.2**   If $f$ is a convex functional defined on $K$, then the set

$$A_\alpha = \{x \in K: f(x) \le \alpha\} \subset K \subset X$$

is convex for each $\alpha \in R$.

$A_\alpha$ is called the level set.

**Proof.**   Let f be convex on $K$. Let $x,\ y \in A_\alpha$ and $0 < \lambda < 1$.
Then

$$f\,(\lambda x + (1 - \lambda)\, y) \ \le\ \lambda f\,(x) + (1 - \lambda)\, f(y)$$

$$\le\ \lambda\alpha + (1 - \lambda)\, \alpha = \alpha.$$

Hence $\lambda x + (1 - \lambda)\, y \in A_\alpha$ and thus $A_\alpha$ is convex.

**Theorem 3.1.3**   Let $f$ be a convex functional defined on a convex subset $K$ of a nls $X$. Let $\mu = \inf\limits_{x \in K} f(x)$. Then

(i) $\Omega = \{x \in K: f(x) = \mu\} \subset K$ is convex

(ii) if $x_0$ is a local minimum of $f$, then $f(x_0) = \mu$ and hence $x_0$ is global minimum.

**Proof.**   (i) Let $x_1,\ x_2 \in \Omega,\ 0 < \lambda < 1$, and $x = \lambda x_1 + (1 - \lambda)\, x_2$.

Then
$$f(x) = f(\lambda x_1 + (1 - \lambda)\, x_2)$$

$$\le \lambda f(x_1) + (1 - \lambda)\, f(x_2)$$

$$\le \lambda\mu + (1 - \lambda)\, \mu = \mu.$$

But for any $x \in K,\ f(x) \ge \mu$.

Thus $f(x) = \mu$, and hence $x \in \Omega$.

Therefore $\Omega$ is a convex set.

(ii) Suppose $N$ is a nghd about $x_0$ in which $x_0$ minimizes $f$.

For any $y \in K$, there is an $x \in N$ such that

$$x = \lambda x_0 + (1 - \lambda)\, y \text{ for some } 0 < \lambda < 1.$$

We have

$$f(x_0) \le f(x) = f(\lambda x_0) + (1 - \lambda)\, f(y))$$

$$\leq \lambda\, f(x_0) + (1 - \lambda)\, f(y)$$

or $$\qquad (1 - \lambda)\, f(x_0) \leq (1 - \lambda)\, f(y)$$

or $$\qquad f(x_0) \leq f(y)$$

and this completes the proof.

**Theorem 3.1.4**  Let $K$ be a convex subset of a linear space $X$. Let $f : K \to R$ and $g : K \to R$ be convex and let $\alpha \geq 0$. Then the functions $f + g$ and $\alpha f$ are convex.

**Theorem 3.1.5**  Let $K$ be a convex subset of a linear space $X$. Let $f : K \to R$ and $g : I \to R$ be such that $\mathrm{Rng} \subset I = \mathrm{Dom}\ g$ where $I$ denotes an interval in $R$. If $f$ and $g$ are convex and $g$ non decreasing, then the composite function go $f$ is convex on $K$.

**Proof.**  Let $x,\ y \in K$ and $0 < \lambda < 1$.

Then $g[f(\lambda x + (1 - \lambda)\, y)]$

$$\leq g[\lambda f(x) + (1 - \lambda)\, f(y)]$$

$$\leq \lambda g(f(x)) + (1 - \lambda)\, g(f(y)),$$

and this completes the proof.


## 3.2  LOWER-SEMI CONTINUOUS CONVEX FUNCTIONALS

We shall first define lower-semi continuous functions and present the properties of their epigraphs and level sets.

Let $X$ be a topological space.

**Definition 3.2.1**  The function $f : X \to R$ is called lower semi continuou at $x_0$ if

$$f(x_0) = \liminf_{x \to x_0} f(x)$$

Similarly $f$ is called upper-semi continuous at $x_0$,
if

$$f(x_0) = \limsup_{x \to x_0} f(x).$$

We recall that,

$$\liminf_{x \to x_0} f(x) = \sup_{V \in V(x_0)}\ \inf_{x \in V} f(x)$$

and

$$\limsup_{x \to x_0} f(x) = \inf_{V \in V(x_0)}\ \sup_{x \in V} f(x)$$

where $V(x_0)$ is a local base at $x_0$.

Hence $f$ is upper-semi continuous if and only if $-f$ is lower-semi continuous, we shall analyse only lower-semi continuous functions.

A function which is lower-semi continuous at each point of $X$ is called lower-semi continuous on $X$.

**Remark 3.2.1**   Consider the topology

$$T_1 = \{(a, \infty): a \in R\} \cap \{R, \Phi\} \text{ on the real line.}$$

Then we see that $f : X \to R$ is lower-semi continuous if and only if $f : X \to (R, T_1)$ is continuous. Thus $f : X \to R$ lower-semi continuous iff $\{x \in X: f(x) \in a\}$, $a \in R$ is closed.

**Theorem 3.2.1**   Let $X$ be a topological space and let $f : X \to R$ be a function. Then $f$ is lower-semi continuous if and only if the epi $f$ is closed in $X \times R$.

**Proof.**   Define $\Phi(x, \alpha) = f(x) - \alpha$, $x \in X$ and $\alpha \in R$. Then $f$ is lower-semi continuous if and only $\Phi$ is lower semi-continuous on $X \times R$ if and only if $\{(x, \alpha): f(x) \leq \alpha\}$ is closed, completes the proof.

**Theorem 3.2.2**   **(Weierstrass):** A lower-semi continuous function on a compact space attains a minimum value.

**Proof.**   If $f$ is lower-semi continuous, each

$$E_\alpha = \{x \in X : f(x) \leq \alpha\}$$

is closed. $E_\alpha \subseteq E_\beta$ iff $\alpha \leq \beta$. Since $X$ is compact, $\underset{\alpha \in R}{\cap} E_\alpha$ is nonempty. Let

$$x_0 \in \underset{\alpha \in R}{\cap} E_\alpha.$$

Clearly $f(x_0) \leq f(x)$ for all $x \in X$.

**Theorem 3.2.3**   If epif is closed, then $f$ is lsc on $K$.

**Proof.**   Observe that the set

$$\{(x, \alpha) \in XXR: x \in X\}$$

is closed for each $\alpha \in R$.

Therefore if epi $f$ is closed, so is

$$epi\ f \cap \{(x, \alpha): x \in X\}$$

$$= \{(x, \alpha): x \in K\ f(x) \leq \alpha\}$$

For each $\alpha \in R$ it now follows that the set

$$T_\alpha = \{x: x \in K, f(x) \leq \alpha\}$$

is closed.

Let $\{x_i\}$ be a sequence in $K$ converging to $x \in K$. Let $b = \lim\limits_{x_i \to x} f(x_i)$. If $b = -\infty$, then

$x \in \overline{T}_\alpha = \overline{T_\alpha}$ for each $\alpha \in R$, which is impossible. Thus $b > -\infty$ and

$$x \in \overline{T}_{b+\epsilon} = T_{b+\epsilon} \quad \text{for all } \epsilon > 0,$$

In other words,

$$f(x) = \leq \lim\limits_{x_i \to x} f(x_i)$$

which shows that $f$ is lsc.

**Theorem 3.2.4**  Let $f$ be a convex functional defined on a normed linear space $X$. Then $f$ is lower-semi continuous if and only if it is weakly lower-semi continuous.

**Proof.**  We know that convex subset of a normed linear space is closed if and only if it is weakly closed. By the above theorem, we see that epi $f$ is closed if and only if it is weakly closed.

**Theorem 3.2.5**  Let $f$ be a lower-semi continuous, convex functional on a reflexive Banach space $X$. Then $f$ takes a minimum value on every closed, bounded and convex subset $K$ of $X$, that is, there exists $x_0 \in K$ such that

$$f(x_0) = \inf \{f(x): x \in K\}.$$

**Proof.**  It is well known that every closed bounded subset of a reflexive space is weakly compact. By the above theorem, $f$ is weakly lower-semi continuous. Now the conclusion follows from the theorem of Weierstrass.

**Theorem 3.2.6**  A convex real function which is finite on a closed interval $[a, b]$ is bounded (from above and below).

**Proof.**  Let $f$ be convex and finite on $[a, b]$. Let $\max\{f(a), f(b)\} = M$. For any

$$z = \lambda a + (1 - \lambda) b \text{ in the interval } [a, b] \text{ we have}$$

$$f(z) = f(\lambda a + (1 - \lambda) b) \leq \lambda f(a) + (1 - \lambda) f(b)$$

$$\leq \lambda M + (1 - \lambda) M = M$$

Thus $f$ is bounded above by $M$,

To see that $f$ is bounded from below, observe that any arbitrary point can be, without any loss of generality, expressed as

$\dfrac{a + b}{2} + t$. Then we have

$$f\left(\frac{a + b}{2}\right) \leq \frac{1}{2} f\left(\frac{a + b}{2} + t\right) + \frac{1}{2} f\left(\frac{a + b}{2} - t\right)$$

Thus

$$f\left(\frac{a+b}{2}+t\right) \geq 2f\left(\frac{a+b}{2}\right) - f\left(\frac{a+b}{2}-t\right)$$

Since $M$ is the upper bound we have

$$-f\left(\frac{a+b}{2}-t\right) \geq -M$$

So

$$f\left(\frac{a+b}{2}-t\right) \geq 2f\left(\frac{a+b}{2}\right) - M = m\,(\text{say}).$$

This completes the proof.

**Definition 3.2.2**  Let $f : [a, b] \to R.$ $f$ is said to be Lipschitzian if there exists a constant $K > 0$ such that

$$|f(x) - f(y)| \leq K\,|x - y|$$

for any two points $x, y \in [a, b]$.

Observe that a Lipschitzian function defined in $[a, b]$ is uniformly continuous on $[a, b]$. For let $f$ be Lipschitzian and let $\epsilon > 0$ be given. Then there exists a $\delta = \epsilon/K$ such that $|x - y| < \delta$ implies $|f(x) - f(y)| < \epsilon$.

**Theorem 3.2.7**  Let $f : I \to R$ be convex. Then $f$ is Lipschitzian on any closed interval $[a, b]$ contained in the interior $I^0$ of $I.f$ is continuous on $I^0$.

**Proof.**  Choose $\epsilon > 0$ in such a way that $[a - \epsilon, b + \epsilon] \subset I$. Let $M$ and $m$ be the upper and lower bounds, respectively, of $f$ on $[a - \epsilon, b + \epsilon]$. If $x, y \in [a, b]$ and $x \neq y$, put

$$z = y + \frac{\epsilon}{|y - x|}\,(y - x), \quad \lambda = \frac{|y - x|}{\epsilon + |y - x|}$$

Then $y = \lambda z + (1 - \lambda)\,x$ and we have

$$f(y) \leq \lambda f(z) + (1 - \lambda)\,f(x)$$

$$= \lambda\,[f(z) - f(x)] + f(x)$$

$$f(y) - f(x) \leq \lambda(M - m) < \frac{|y - x|}{\epsilon}\,(M - m)$$

$$= K|y - x|$$

where $K = \dfrac{M - m}{\epsilon}$. Since this is true for any $x, y \in [a, b]$ we conclude that $f$ is Lipschitzian.

Since $[a, b] \subset I^0$ is an arbitrary closed interval in $I^0$, it follows that $f$ is continuous in $I^0$ and this completes the proof.

**Theorem 3.2.8**  Let $f : I \to R$ be convex. Then for $a, b, x \in I$ and $x \in (a, b)$,

$$\frac{f(x) - f(a)}{x - a} \leq \frac{f(b) - f(a)}{b - a} \leq \frac{f(b) - f(x)}{b - x}$$

If $f$ is strictly convex, then strict inequality holds.

**Proof.**  Since $f$ is convex we have

$$f(x) \leq \frac{b - x}{b - a} f(a) + \frac{x - a}{b - a} f(b).$$

Therefore it follows that

$$f(x) - f(a) \leq \frac{a - x}{b - a} f(a) + \frac{x - a}{b - a} f(b)$$

$$= \frac{x - a}{b - a} \, [f(b) - f(a)]$$

This shows that the first inequality holds. The second one can be proved similarly. Also observe that if $f$ is strictly convex, the above inequality is strict and hence the required inequality follows.

**Theorem 3.2.9**  Let $f : I \to R$ be convex. Then $f$ has both a right and a left derivative, at every point of $I^0$, the interior of $I$, and $f'_-, f'_+$ are non decreasing on $I^0$. If $c \in I^0$, then we have

$$f'_-(c) \leq f'_+(c),$$

$$f(x) \geq f(c) + f'_-(c) \, (x - c),$$

$$f(x) \geq f(c) + f'_+(c) \, (x - c)$$

for all $x \in I$.

**Proof.**  Let $c \in I^0$ and let $[a, b] \subset I$ be such that $c \in (a, b)$. Then we have by the above theorem

$$\frac{f(c) - f(a)}{c - a} \leq \frac{f(x) - f(c)}{x - c}$$

for $x \in (c, b]$. It also follows from the above theorem that the function

$$x \to \frac{f(x) - f(c)}{x - c}$$

is nondecreasing on $(c, b]$. Hence the right derivative

$$f'_+(c) = \lim_{x \to c+} \frac{f(x) - f(c)}{x - c}$$

exists. In a similar manner it can be shown that the left derivative also exists. If $a < c < d < b$ then for sufficiently small positive $h$ we have

$$\frac{f(c) - f(c-h)}{h} \leq \frac{f(c+h) - f(c)}{h} \leq \frac{f(d) - f(d-h)}{h}$$

Letting $h \to 0$ we obtain

$$f'_-(c) \leq f'_+(c) \leq f'_-(d).$$

**Remark 3.2.2** Using the above theorem we now give an alternative and simple proof of the fact that if $f : I \to R$ is convex, then it is Lipschitzian on any closed interval $[a, b] \subset I^0$.

For this choose the points $c$ and $d$ in $I$ such that $c < a < b < d$ and this choice is possible. Then by the above theorem we have

$$f'_+(a) \leq f'_+(x) \leq \frac{f(x) - f(y)}{x - y} \leq f'_-(y) \leq f'_-(b)$$

whenever $a \leq x < y \leq b$. It now follows that

$$|f(x) - f(y)| \leq K\,|x - y|$$

where

$$K = \max\ \{|f'_+(a)|, f'_-(b)\}$$

and this completes the proof.

**Theorem 3.2.10** Let $f : I \to R$ be convex. Then

   (i) $f'_-$ is left continuous and $f'_+$ is right continuous on $I^0$,

   (ii) the set of all points where $f$ is not differentiable is atmost countable.

**Proof.** (i) Since $f : I \to R$ is convex, it is continuous $I^0$. Therefore for all $x, y, z \in I^0$ we have

$$\frac{f(y) - f(x)}{y - x} = \lim_{z \to x+} \frac{f(y) - f(z)}{y - z} \geq \lim_{z \to x+} f'_+(z)$$

$x < z < y$. Letting $y \to x +$ we obtain

$$f'_+(x) \geq \lim_{z \to x+} f'_+(z).$$

Since $f'_+$ is nondecreasing, we have

$$f'_+(x) \leq \lim_{z \to x+} f'_+(z)$$

Thus it follows that

$$f'_+(x) = \lim_{z \to x+} f'_+(z).$$

This shows that $f'_+$ is right continuous.

In a similar manner it can be shown that $f'_-$ is left continuous.

(ii) Let $x$, $y$, $z \in I$ and let $x < y < z$. Then it follows that

$$f'_+(x) \le f'_-(y) \le f'_+(z)$$

Suppose that $f'_+$ is continuous at $y$. Then

$$f'_+(y) = \lim_{x \to y^-} f'_-(x) = \lim_{z \to y^+} f'_+(z) = f'_-(z)$$

This shows that $f$ is differentiable at $y$. Thus the points of $I^0$ where $f$ is not differentiable are these points where the nondecreasing function $f'_+$ has a jump and there are only countabley many such jumps. This completes the proof of (ii).

**Theorem 3.2.11**   Let $I$ be open and let $f : I \to R$ be twice differentiable. Then $f$ is convex if and only if $f''(x) \ge 0$ for all $x \in I$.

**Proof.**   Let $f$ be convex. Then since $I$ is open, $f'$ is nondecreasing on $I$. Hence $f''(x) \ge 0$ for all $x \in I$.

Conversely let $f''(x) \ge 0$ for all $x \in I$. Let $x$, $y \in I$, $x < y$ and $0 < \lambda < 1$. Then by Mean-Value Theorem for differential calculus there exist $\xi_1$, $\xi_2$, $x < \xi_1 < \lambda x + (1 - \lambda) y < \xi_2 < y$ and there exists $\xi_3$ with $\xi_1 < \xi_3 < \xi_2$ such that

$$f(\lambda x + (1 - \lambda) y) - \lambda f(x) - (1 - \lambda) f(y)$$
$$= \lambda \, [f(\lambda x + (1 - \lambda)y) - f(x)] + (1 - \lambda) \, [f(\lambda x + (1 - \lambda) y) - f(y)]$$
$$= \lambda (1 - \lambda) (y - x) f'(\xi_1) + (1 - \lambda) \lambda) - f(x - y) f'(\xi_2)$$
$$= \lambda (1 - \lambda) (y - x) (\xi_1 + \xi_2) f''(\xi_3) \le 0$$

Thus $f$ is convex.

It should be noted that if $f$ is strictly convex, then $f''(x) > 0$ for all $x \in I$, but the converse is not always true. For example, the function $f$ defined by

$$f(x) = x^4$$

is strictly convex on $R$ but $f''(o) = 0$

**Theorem 3.2.12**   Let the convex functional $f$ be Gateaux differentiable at $x_0 \in$ int $D(f)$ with the Gateaux derivative $df(x_0)$. Then

$$f(x) \ge f(x_0) + (df(x_0), x - x_0)$$

for all $x \in X$.

**Proof.**　Since $f$ is convex, we have for any $x \in X$ and $0 < \lambda < 1$,

$$f(x_0 + \lambda \, (x - x_0)) - f(x_0)$$
$$= f(\lambda x + (1 - \lambda) \, x_0) - f(x_0)$$
$$\leq \lambda f(x) + (1 - \lambda) \, f(x_0) - f(x_0)$$
$$= \lambda \, [f(x) - f(x_0)]$$

Hence

$$f(x) - f(x_0) \geq \lim_{\lambda \to 0+} \frac{f(x_0 + \lambda \, (x - x_0)) - f(x_0)}{\lambda}$$

$$= (df(x_0), \, x - x_0).$$

## 3.3　SUBGRADIENT, SUBDIFFERENTIABILITY

The above theorem leads to the following definition in a natural way.

**Definition 3.3.1**　An element $x_0^* \in X^*$ is said to be subgradient of $f$ at $x_0$ if

$$f(x) \geq f(x_0) + \left( x_0^*, \, x - x_0 \right)$$

for all $x \in X$.

The set of all subgradients of $f$ at $x_0$ is denoted by $\partial f(x_0)$.

Now we have multi-valued map

$$\partial f \colon X \to 2^{x^*}$$

where $\partial f(x)$ is the set of all subgradients of $f$ at $x$.

This map $\partial f$ is called the subdifferential of $f$ at $x$ and if $\partial f(x) \neq \Phi$, then we say that $f$ is subdifferentiable at $x$.

**Theorem 3.3.1**　If $f$ is every where Gateaux differentiable, then $\partial f$ is single-valued and $\partial f = \Delta f \colon X \to X^*$.

**Proof.**　It is enough to show that $\partial f(x_0) \subset \{\nabla f(x_0)\}$, i.e., if $x_0^* \in \partial f(x_0)$, then for all $h \in X$,

$$\left( x_0^*, \, h \right) = (\Delta f(x_0), \, h)$$

by the definition　for $t > 0$,

$$\frac{\left( x_0^*, \, th \right)}{t} \leq \frac{f(x_0 + th) - f(x_0)}{t}$$

Taking limit as $t \to 0$ we get

$$\left(x_0^*, h\right) = \frac{(x_0^*, th)}{t}$$

$$\leq \lim_{t \to 0+} \left[\frac{f(x_0 + th) - f(x_0)}{t}\right]$$

$$= (\nabla f(x_0), h).$$

Now replacing $t$ by $-t$ we get

$$\left(x_0^*, h\right) \geq (\nabla f(x_0), h)$$

and this completes the proof.


## 3.4   DUALITY MAPPING

**Definition 3.4.1**   A mapping $J: X \to 2^{X^*}$ is called the duality mapping if

$$Jx = \{x^* \in X^*: \left(x^*, x\right) = \|x\| \, \|x^*\|, \, \|x\| = \|x^*\|\}$$

In general $J$ is a multi-valued map and we shall observe in a later chapter that it is single-valued if $X^*$ is strictly convex.

When $J$ is single-valued we have

$$(Jx, x) = \|Jx\| \, |x\|, \, \|Jx\| = \|x\|.$$

A duality map can be constructed in a nls in the following way:

It follows from Hahn-Banach theorem that for any $x \in X$, $\exists$ at least one bounded linear functional $y_x \in X^*$ such that

$$\|y_x\| = 1, \, (y_x, x) = \|x\|.$$

Let $y_x$ be one such functional and set

$$Jx = \|x\| \, y_x, \, J(-x) = -\|x\| \, y_x.$$

We get

$$Jx = \|x\|, \, (Jx, x) = \|Jx\| \, \|x\|.$$

**Theorem 3.4.1**   The subdifferential of the norm functional $j$ defined by

$$j(x) = \frac{\|x\|^2}{2}$$

is precisely the duality mapping $j$.

**Proof.**   By definition

$$\partial j(x_0) = \left\{x_0^* : j(x) \geq j(x_0) + (x_0^*, x - x_0) \, \nabla x \in X\right\}$$

Let $\epsilon > 0$ be arbitrary. Put $x = (1 \pm \epsilon)x_0$. Then for $x_0^* \in \partial j(x_0)$ we have

$$\frac{1}{2}(1 \pm \epsilon)^2 \|x_0\|^2 \geq \frac{1}{2}\|x_0\|^2 \pm \epsilon\left(x_0^*, x\right).$$

That is

$$\frac{1}{2}(1 + \epsilon^2 \pm 2\epsilon)\|x_0\|^2 \frac{1}{2}\|x^0\|^2 \pm \epsilon(x_0^*, x)$$

or

$$\pm \epsilon\left(x_0^*, x\right) \leq \pm \epsilon\|x_0\|^2 \pm \frac{1}{2}\epsilon^2\|x_0\|^2$$

This gives

$$\|x^0\|^2 - \frac{\epsilon}{2}\|x_0\|^2 \leq \left(x_0^*, x\right)$$

$$\leq \|x^0\|^2 + \frac{\epsilon}{2}\|x_0\|^2$$

Letting $\epsilon \to 0$ we get

$$\left(x_0^*, x\right) = \|x_0\|^2$$

and this gives

$$\|x_0^*\| \geq \|x_0\|. \qquad \qquad ...(3.1)$$

For $\|x_0\| = 1$, using the definition of $\partial j(x_0)$ we get that for

$$x_0^* \in \partial j(x_0),$$

$$(x_0^*, x) - \|x_0\|^2 \leq \frac{1}{2}[\|x\|^2 - \|x_0\|^2]$$

This gives $(x_0^*, x) \leq 1 \forall x$ with $\|x\| = 1$ and then $\|x_0^*\| \leq 1$ for $\|x_0\| = 1$.
Now, since $\partial j$ is a homogeneous operator we have

$$\partial j(x_0) = \|x_0\| \, \partial j\left(\frac{x_0}{\|x_0\|}\right) \qquad \qquad ...(3.2)$$

and hence we get the inequality $\|x_0^*\| \leq \|x_0\|$ for arbitrary $x_0^* \in \partial j(x_0)$.
Combining (3.1) and (3.2) we get that for $x_0^* \in \partial j(x_0)$,

$$\|x_0^*\| = \|x_0\|^2.$$

We have already proved that

$$\left(x_0^*, x_0\right) = \|x_0\|^2.$$

So we have

$$\partial j(x_0) = \{x_0^* : (x_0^*, x_0) = \|x_0\|^2, \|x_0^*\| = \|x_0\|\}$$

$$= J(x_0).$$

## 3.5  CONJUGATE FUNCTIONALS

**Definition 3.5.1**  Let $f$ be a convex functional defined on a convex set $K$ in a nls $X$. The conjugate set $K^*$ is defined by

$$K^* = \{x^* \in X^*: \sup_{x \in K} \left[(x^*, x) - f(x)\right] < \infty \}.$$

The conjugate functional $f^*$ is defined on $K^*$ as follows :

$$f^*(x^*) = \sup_{x \in K} \, [(x^*, x) - f(x)].$$

**Example 3.5.1**  If $X = R^n$, and for $x = (x_1, x_2, \ldots x_n)$,

$$f(x) = \frac{1}{p} \sum_{i=1}^{n} |x_i|^p, \; 1 < p < \infty,$$

then

$$f^*(x^*) = \frac{1}{q} \sum_{i=1}^{n} |x_i^*|^q$$

where $x^* = (x_1^*, x_2^*, \ldots x_n^*)$ and $\frac{1}{p} + \frac{1}{q} = 1$.

**Proof.**  We have

$$f^*(x^*) = \sup \left[(x, x^*) - \frac{1}{p} \sum_{i=1}^{n} |x_i|^p\right]$$

$$= \sup \left[\sum_{i=1}^{n} x_i^* \, x_i - \frac{1}{p} \sum_{i=1}^{n} |x_i|^p\right]$$

$$= \sum_{i=1}^{n} |x_i|^p - \frac{1}{p} \sum_{i=1}^{n} |x_i|^p$$

(Since the supremum is attained at $x_i^*) = |x_i|^{p-1} \, \text{sgn} \, x_i$

$$= \left(1 - \frac{1}{p}\right) \sum_{i=1}^{n} |x_i|^p$$

$$= \frac{1}{q} \sum_{i=1}^{n} |x_i^*|^q$$

**Theorem 3.5.1**  The conjugate set $K^*$ and the conjugate functional $f^*$ are convex and epi $f^*$ is a closed convex subset of $R \times X^*$.

**Proof.**  For any $x_1^*, x_2^* \in X^*$ and $\lambda$, $0 < \lambda < 1$ we have

$$\sup_{x \in K} \, [(x, \lambda x_1^* + (1 - \lambda) \, x_2^*) - f(x)]$$

$$= \lambda \sup_{x \in K} [(x, x_1^*) - f(x)] + (1 - \lambda) \sup_{x \in K} [(x, x_2^*) - f(x)]$$

It now follows that $K^*$ and $f^*$ are convex.

We now show that epi $f^*$ is closed. To that end, let $\{(s_i, x_i^*)\}$ be a convergent sequence from epi $f^*$ with $(s_i, x_i^*) \to (s, x^*)$. We claim that $(s, x^*) \in$ epi $f^*$. To establish the claim, observe that for every $i$ and every $x \in K$, we have

$$s_i \geq f^*(x_i^*) \geq (x, x_i^*) - f(x)$$

Letting $i \to \infty$ we obtain for all $x \in K$,

$$s \geq (x, x^*) - f(x)$$

Therefore

$$s \geq \sup_{x \in K} \lfloor (x, x^*) - f(x) \rfloor$$

It now follows that

$$x^* \in K^* \text{ and } s \geq f^*(x^*),$$

and this completes the proof.

## 3.6  QUASI CONVEX AND PSEUDO CONVEX FUNCTIONS

In this section we shall discuss about certain important generalized convex functions – namely the quasi – and the pseudo-convex functions. These are most commonly used non-convex functions but are close to convex in some sense. Although convexity plays a vital role in applications, but most real world problems are non-convex in nature. Therefore the quasi-and the pseudo-convex functions have found applications in Economics and Operations Research and hence the study of such functions has become essential in present days.

We shall first discuss certain basic and important properties of differentiable convex and concave functions. Subsequently we shall discuss about quasi-convex and finally about pseudo-convex functions.

**Theorem 3.6.1**  Let $f$ be a real-valued function defined on an open set $G \subset R^n$ and let $f$ be differentiable at a point $\bar{x} \in G$. If $f$ is convex at $\bar{x} \in G$, then

$$f(x) - f(\bar{x}) \geq \nabla f(\bar{x}) (x - \bar{x}) \text{ for cach } x \in G.$$

If $f$ is concave at $\bar{x} \in G$, then

$$f(x) - f(\bar{x}) \leq \nabla f(\bar{x}) (x - \bar{x}) \text{ for each } x \in G.$$

**Theorem 3.6.2**  Let $f$ be a real differentiable function on an open convex set $K \subset R^n$. $f$ is convex on $K$ iff

$$f(x_2) - f(x_1) \geq \nabla f(x_1) \, (x_2 - x_1) \text{ for all } x_1, x_2 \in K \text{ and } f \text{ is concave on } K \text{ iff}$$

$$f(x_2) - f(x_1) \leq \nabla f(x_1) \, (x_2 - x_1) \text{ for all } x_1, x_2 \in K.$$

**Proof.**   We shall prove only for the convex case. The proof for the concave case is similar.

Necessity. Since $f$ is convex at $x_1 \in K$, this follows from the previous theorem.

Sufficiency. Let $x_1, x_2 \in K$ and $0 \leq \lambda \leq 1$. Since $K$ is convex, $(1 - \lambda) \, x_1 + \lambda x_2 \in K$.

We have

$$f(x_1) - f\lfloor (1 - \lambda) \, x_1 + \lambda x_2 \rfloor \geq \lambda \nabla f\lfloor (1 - \lambda) \, x_1 + \lambda x_2 \rfloor \, (x_1 - x_2)$$

Also

$$f(x_2) - f\lfloor (1 - \lambda) \, x_1 + \lambda x_2 \rfloor \geq - (1 - \lambda) \, \nabla f\lfloor (1 - \lambda) \, x_1 + \lambda x_2 \rfloor \, (x_1 - x_2).$$

Multiplying the first inequality by $(1 - \lambda)$, the second by $\lambda$ and adding we get.

$$(1 - \lambda) \, f(x_1) + \lambda f(x_2) \geq f\lfloor (1 - \lambda) \, x_1 + \lambda x_2 \rfloor$$

and this completes the proof.

**Theorem 3.6.3**   Let $f$ be a real differentiable function on an open convex set $K \subset R^n$. A necessary and sufficient condition that $f$ is convex (concave) on $K$ is that for each $x_1, x_2 \in K$,

$$\lfloor \nabla f(x_2) - \nabla f(x_1) \rfloor \, (x_2 - x_1) \geq 0 \, (\leq 0).$$

We now introduce the definitions of Quasi-convex and Quasi-concave functions.

A real – valued function defined on a set $K \in R^n$ is said to be Quasi-convex at $\bar{x} \in K$ if

$$x \in K, f(x) \leq f(\bar{x}), 0 \leq \lambda \leq 1, (1 - \lambda) \, \bar{x} + \lambda x \in K$$

$$\Rightarrow \ f\lfloor (1 - \lambda) \, \bar{x} + \lambda x \rfloor \leq f(\bar{x}).$$

$f$ is said to be Quasi-convex on $K$ iff

$$x_1, x_2 \in K, f(x_2) \leq f(x_1), 0 \leq \lambda \leq 1$$

$$\Rightarrow f\lfloor (1 - \lambda) \, x_1 + \lambda x_2 \rfloor \leq f(x_1).$$

$f$ defined on a set $K \in R^n$ is Quasi-concave at $\bar{t} \in K$ if

$$x \in K, f(\bar{x}) \leq f(x), 0 \leq \lambda \leq 1, (1 - \lambda) \, \bar{x} + \lambda x \in K$$

$$\Rightarrow \qquad f(\bar{x}) \le f\lfloor (1 - \lambda)\,\bar{x} + \lambda x \rfloor$$

$f$ defined on a convex set $K \subset R^n$ is Quasi-concave on $K$ if

$$x_1,\, x_2 \in K,\, f(x_1) \le f(x_2),\, 0 \le \lambda \le 1$$

$$\Rightarrow \qquad f(x_1) \le f\lfloor (1 - \lambda)\,x_1 + \lambda x_2 \rfloor$$

$f$ is said to be strictly Quasi-convex (strictly Quasi-concave) if strict inequality holds in each of the above cases respectively.

**Theorem 3.6.4**  Let $f$ be a real function defined on a convex set $K \in R^n$.

For any $\alpha > 0$ let

$$\Lambda_\alpha = \{x \in K : f(x) \le \alpha\}$$

and $\qquad \Omega_\alpha = \{x \in K : f(x) \le \alpha\}$

Then $f$ is quasi-convex on $K$ iff $\Lambda_\alpha$ is a convex set for every $\alpha > 0$. $f$ is quasi-concave on $K$ iff $\Omega_\alpha$ is convex for every $\alpha > 0$.

**Proof.**  Let $\Lambda_\alpha$ be convex for each $\alpha > 0$. Let $x_1,\, x_2 \in K,\, f(x_2) \le f(x_1),\, 0 \le \lambda \le 1$. Let $\alpha = f(x_1)$. By the convexity of $\Lambda_\alpha$,

$$f\lfloor (1 - \lambda)\,x_1 + \lambda x_2 \rfloor \le \alpha = f(x_1).$$

Thus $f$ is quasi-convex.

Conversely, let $f$ be quasi-convex, $\alpha$ any real number, $x_1,\, x_2 \in \Lambda_\alpha$. Without any loss of generality, we may assume, that $f(x_2) \le f(x_1)$. Since $x_1,\, x_2 \in \Lambda_\alpha$,

$$f(x_2) \le f(x_1) \le \alpha.$$

Since $f$ is quasi-convex and $K$ is convex, for $0 \le \lambda \le 1$ we have

$$f\lfloor (1 - \lambda)\,x_1 + \lambda x_2 \rfloor \le f(x_1) \le \alpha.$$

Hence $(1 - \lambda)\,x_1 + \lambda x_2 \in \Lambda_\alpha$. Thus $\Lambda_\alpha$ is convex and this completes the proof.

*Remark 3.6.1*  It should be noted that a strictly quasi-convex function need not be quasi-convex. For example, let $f : R \to R$ be defined by

$$f(x) = \begin{cases} 1, & x = 0 \\ 0, & x \ne 0 \end{cases}$$

Then $f$ is strictly quasi-convex, but it is not quasi-convex on $R$. However a lower semi-continuous strictly quasi-convex function is quasi-convex.

# Invex and Related Functions

4

In this chapter we shall discuss a generalization of convex functions, called the invex functions. This is of recent origin introduced by Hanson in the year 1981. Recall that a differentiable function f is convex if and only if

$$f(y) - f(x) \geq (\nabla f(x), y - x)$$

for all $x, y \in R^n$. Hanson observed that the term $(y - x)$ on the right hand side of the above inequality plays no role in the proof on the sufficiency of the Kuhn – Tucker conditions. This motivated Hanson to introduce, a class of differentiable functions called invex. This is introduced with reference to a vector function. We now quote the definition below.

**Definition 4.1.1**  Let $f : R^n \to R$ be a differentiable function and $\eta : R^n \times R^n \to R$ be an arbitrary function. $f$ is said to be invex (with respect to $\eta$) if

$$f(y) - f(x) \geq (\nabla f(x), \eta(y, x))$$

for all $x, y \in R^n$. Note that if $\eta(y, x) = y - x$, then $f$ becomes a convex function. After introducing this concept Hanson proved that Kuhn-Tucker conditions are also sufficient if the objective and constraint functions are invex with respect to the same function $\eta$.

We now introduce the definitions of two more concepts, which are generalizations of invex functions.

**Definition 4.1.2**  $f$ is said to be pseudo-invex with respect to $\eta : R^n \times R^n \to R^n$ if

$$(\nabla f(x), \eta(y, x)) \geq 0 \Rightarrow f(y) \geq f(x).$$

**Definition 4.1.3**  $f$ is said to be quasi-invex with respect to $\eta$ if

$$f(y) \geq f(x) \Rightarrow (\nabla f(x), \eta(y, x)) \geq 0$$

We now discuss some properties.

**Theorem 4.1.1**  Every convex function is invex, but the converse is not true.

**Proof.**  Let $f$ be a differentiable convex function. Then

$$f(y) - f(x) \geq (\nabla f(x),\ y - x)$$

for all $x, y \in R^n$. Therefore the inequality required for the function to be invex is satisfied for $\eta(y, x) = y - x$ and hence $f$ is invex.

To show that the converse is not true, consider the function $f:\left[0,\ \dfrac{\pi}{2}\right] \to R$ defined by

$$\eta(y, x) = \frac{\sin y - \sin x}{\cos x}$$

This is not convex, can be seen by taking $x = \dfrac{\pi}{4}$ and $y = \dfrac{\pi}{6}$.

**Theorem 4.1.2**  Every quasi-convex function is quasi-invex, but not conversely.

**Proof.**  If $f$ is differentiable quasi-convex, then the inequality required for $f$ to be quasi-invex is satisfied for $\eta(y, x) = y - x$.

For the converse part, let $f : [0, \pi] \to R$, be given by

$$f(x) = \sin^3 x.$$

This is quasi-invex for

$$\eta(y, x) = \cos x(\sin y - \sin x).$$

This is not quasi-convex, for if

$$x = \frac{\pi}{4}, \qquad\qquad y = \frac{3\pi}{4}, \text{ then}$$

$$f(x) = f(y), \qquad\qquad \text{but } (y - x)t\ \nabla f(x) > 0.$$

**Theorem 4.1.3**  If $f$ is pseudo-convex, then it is pseudo-invex, but the converse is not true.

**Proof.**  Observe that if $f$ is pseudo-convex, then it is pseudo-invex with

$$\eta(y, x) = y - x. \text{ Let}$$

$$C = \{(x, y) \in R^2 : 4x^2 + 4y^2 - 9 \leq 0,\ x, y \geq 0\}.$$

Let $f : C \to R$ be given by

$$f(x, y) = x + \sin y.$$

This is pseudo-invex for

$$\eta(x,y),(u,v)) = \begin{pmatrix} \dfrac{\sin x - \sin u}{\cos u} \\ \dfrac{\sin y - \sin v}{\cos v} \end{pmatrix}$$

but it is not pseudo-convex

**Theorem 4.1.4**  If $f$ is invex, then it is quasi-invex, but the converse is not true.

**Proof.**  It is obvious from the definition, that invexity implies quasi-invexity. For the converse part consider the function $f : R \to R$ gives by $f(x) = x^3$.

**Theorem 4.1.5**  If $f$ is invex, it is pseudo-invex, but not conversely.

**Proof.**  It is obvious that invex function is Pseudo-invex. For the converse, part consider

$$f : \left( -\frac{\pi}{2}, \frac{\pi}{2} \right) \to R$$

$$f(x) = -\cos^2 x.$$

This is pseudo-invex for

$$\eta\ (y,\ x)\ =\ \sin x\ (\cos y - \cos x)$$

but not invex for $\quad y = -\dfrac{\pi}{6},\ x = -\dfrac{\pi}{4.}$

**Theorem 4.1.6**  There exist invex functions which are not qausi-convex and there exist quasi-convex functions which are not invex.

**Proof.**  Observe that the function $f(x) = x^3$ is quasi-convex, but is not invex for any $\eta$. Similarly the function $f : R^2 \to R$ given by

$$f(x,\ y) = -x^2 + xy - e^y$$

is invex, but it is not quasi-convex.

**Theorem 4.1.7**  Let $\varphi : R \to R$ be a monotone increasing differentiable convex function. If $f$ is invex, then the composite function $\varphi of$ is invex.

**Proof.**  Since $\varphi(x + h) \geq \varphi(x) + \varphi'(x)\ h$ for every $x, \in R$ we get

$$\varphi(f(x))\ \geq\ \varphi(f(y) + f(y)\eta\ (x,y))$$

$$\geq\ \varphi(f(y)) + \varphi(f(y))\nabla\ f(y)\eta(x,y)$$

$$=\ \varphi(f(y)) + \nabla(\varphi of)(y)\ \eta(x,y)$$

This completes the proof.

**Theorem 4.1.8**  If $f$ is a differentiable function and there exists a sequence $\{t_n\}$ of positive real numbers such that $t_n \to 0$ as $n \to \infty$ and

$$f(y + t_n \, \eta(x, y)) \leq t_n f(x) + (1 - t_n) \, f(y) \; (x) \qquad\qquad \dots(4.1)$$

for all $x$, $y$ on the domain of $f$, then $f$ is invex.

**Proof.** Because of (4.1) we get

$$\frac{f(y + t_n \eta \,(x,y)) - f(y)}{t_n} < f(x) - f(y)$$

and this completes the proof.

Note that invex functions are defined for differentiable functions. We now discuss some generalizations of invexity to nondifferentiable case. For this purpose we introduce a generalization of convex sets called invex.

**Definition 4.1.4**    A set $\Lambda \subset R^n$ is called invex with respect to a given

$$\eta : R^n \times R^n \to R^n \text{ if}$$

$$x, y, \in \Lambda, \, 0 \leq \lambda \leq 1 \Rightarrow y + \lambda\eta \, (x, y) \in \Lambda.$$

**Definition 4.1.5**    Let $A \subset R^n$ be an invex set and
$\eta : R^n \times R^n \to R^n$ . $f : A \to R$ is said to be pre-invex with respect to $\eta$ if

$$x, y \in A, \, 0 \leq \lambda \leq 1 \quad \Rightarrow$$

$$(y + \lambda\eta(x, y)) \leq \lambda(x) + (1 - \lambda)f(y).$$

**Definition 4.1.6**    Let $\Lambda \subset R^n$ be an invex set. $f : \Lambda \to R$ is said to be prequasi invex with respect to $\eta : R^n \times R^n \to R$ if

$$x, y \in \Lambda, \, 0 \leq \lambda \leq 1 \quad \Rightarrow$$

$$f(y + \lambda\eta(x, y)) \leq \max \, (f(x), f(y)).$$

### Theorem 4.1.9

(a) If $f$ is differentiable and pre invex (with respect to $\eta$), then it is invex (with respect to same $\eta$).

(b) If $f$ is differentiable and invex, it need not be pre-invex. Infact pre-invexity is a stronger condition than invexity.

**Proof.** (a) This is exactly Theorem 4.1.8.

(b) Consider the function $f : R \to R$ given by $f(x) = x^2$. This is invex with respect to $\eta$ given by

$$\eta(x, y) = \begin{cases} \dfrac{x^2 - y^2}{2y}, & y \neq 0 \\ 0 & y = 0 \end{cases}$$

An easy computation shows that the inequality

$$f(y + t\eta(x, y)) \le tf(x) + (1 - t) f(y)$$

holds if and only if

$$t^2(x^4 - 2x^2y^2 + y^2) \le 0,$$

That is if $t = 0$ or $x = y = 0$.

This completes the proof.

It is now a natural question to ask : when a differentiable invex function will be pre-invex? This was settled by Mohan and Neogy to be true under a condition called condition $C$.

**Condition C:** Let $\eta : R^n \times R^n \to R^n$ We say that $\eta$ satisfies condition $C$ if for any $x, y,$

$$\eta(y, x + \lambda\eta(x, y)) = - \lambda\eta(x, y)$$

$$\eta(x, y + \lambda\eta(x, y)) = (1 -\lambda)\eta(x, y)$$

for all $0 \le \lambda \le 1$.

**Theorem 4.1.10** Suppose that $\Lambda$ is a pre-invex set with respect to $\eta$ and $f : X \to R$ is differentiable where $X$ is open and $\Lambda \subset X$. Further suppose that $f$ is invex with respect to $\eta$ on $\Lambda$ and that $\eta$ satisfies condition $C$. Then $f$ is pre-invex with respect to $\eta$ on $\Lambda$.

**Proof.** Suppose that $x, y \in \Lambda$. Let $0 < \lambda < 1$ and $\overline{X} = y + \lambda\eta(x, y)$. Note that $x \in A$. By the invexity of $f$ we have

$$f(x) - f(\overline{x}) \le (\nabla f(x), \eta(x, \overline{x})) \qquad \qquad \dots(4.2)$$

Similarly the invexity condition applied to the pair $y$ and $\overline{x}$ yields,

$$f(y) - f(\overline{x}) \ge (\nabla f(\overline{x}), \eta(y, \overline{x})) \qquad \qquad \dots(4.3)$$

Multiplying (4.2) by $\lambda$ and (4.3) by $(1 - \lambda)$ and adding we get

$$\lambda f(x) +(1 - \lambda) f(y) - f(\overline{x})$$

$$\ge (\nabla f(x), \lambda\eta(x, \overline{x}) + (1 - \lambda) \eta(y, \overline{x}))$$

By condition $C$,

$$\lambda\eta(x, \overline{x}) + (1 - \lambda) + \eta(y, \overline{x}) = 0$$

Hence the conclusion of the theorem holds.

We shall now discuss generalization of invexity for non-differentiable functions. In other words, it is called semi-invex. For this purpose, let us first quote the definition of invexity for differentiable functions.

A function $f : R^n \to R$ is said to be invex with respect to $\eta : R^n \times R^n \to R^n$ if

$$f(y) - f(x) \geq (\nabla f(x), \eta(y, x))$$

Observe that $\nabla f(x)$ is not the only element in $R^n$ which satisfies the above inequality with the given $\eta$. For example if $f : R \to R$ is a constant function and $\eta(y, x) = (y - x)^2$, then $x^x \in R$ such that $x^x \leq 0$, satisfies the following inequality

$$f(y) - f(x) \geq (x^x, \eta(y, x)).$$

This was the motivation for introduction semi-invex functions. We now quote the definition.

**Definition 4.1.7**   A function $f : R \to R$ is said to be semi-invex at $x \in R^n$ with respect to a given $\eta : R^n \times R^n \to R^n$ if there exists $\xi \in R^n$ such that for all $y \in R^n$,

$$f(y) - f(x) \geq (\xi, \eta(y, x))$$

A function is said to be semi-invex if it is semi-invex at each $x \in R^n$.

**Theorem 4.1.11**   Let $f : R^n \to R$ be semi-invex and let condition $C$ be satisfied. Then $f$ is pre-invex.

**Proof.**   Since $f$ is semi-invex, we have

$$f(y) \geq f(x + \lambda\eta(y, x)) + (\xi, \eta(y, x + \lambda\eta(y, x)))$$

$$f(x) \geq f(x + \lambda\eta(y, x)) + \xi, \eta(x, x + \lambda\eta(y, x))$$

Since $\eta$ satisfies condition $C$ we have

$$f(y) \geq f(x + \lambda\eta(y, x)) + (1 - \lambda)(\xi, \eta(y, x)) \qquad \qquad ...(4.4)$$

$$f(x) \geq f(x + \lambda\eta)) - \lambda(\xi, \eta(y, x)) \qquad \qquad ...(4.5)$$

Multiplying $\lambda$ in (4.4), $(1 - \lambda)$ in (4.5) and adding we get

$$f(x + \lambda\eta(y, x)) \leq \lambda f(y) + (1 - \lambda) f(x)$$

This completes the proof.

**Theorem 4.1.12**   Let $f : R^n \to R$ and $\eta : R^n \times R^n \to R^n$ be such that

(i) $\qquad \qquad \qquad f$ is pre-invex

(ii) $\qquad \qquad \qquad \sup_t \dfrac{f(x + t\eta(y, x)) - f(x)}{t} \geq (\xi, \eta(y, x))$

For some $\xi \in R^n$ and $t \in [0, 1]$. Then $f$ is semi-invex.

**Proof.**   Since $f$ is pre-invex we have for $t \in [0, 1]$,

$$tf(y) + (1 - t) f(x) \geq f(x + t\eta(y,x))$$

So

$$f(y) - f(x) \geq \frac{f(x + t \, \eta \, (y, \, x)) - f(x)}{t}$$

Thus

$$f(y) - f(x) \geq \sup_{t} \frac{f(x + t\eta(y, \, x)) - f(x)}{t}$$

$$\geq (\xi, \, \eta(y, \, x))$$

Thus $f$ is semi-invex and this completes the proof.

## 4.1   LOGARITHIC AND HARMONIC CONVEX FUNCTIONS

In this section we shall discuss the concepts of Logarithmic and Harmonic convexities. Convexity, Logarithmic convexity and Harmonic convexity are related to each other in the same manner as arithmetic mean is related to geometric and harmonic mean. Harmonic convexity is of fairly recent origin, introduced by C. Das in the year 1975. We first start with definitions and then discuss some of their basic properties.

**Definition 4.1.8**   A positive real function $f$ defined on a convex set $K \subset R^n$ is said to be Logarithmic convex (L-convex) if for each $x_1$, $x_2 \in K$ and $0 < \lambda < 1$,

$$f(1 - \lambda)x_1 + \lambda x_2) \leq (f(x_1)^{1-\lambda} \, (f(x_2))^{\lambda}$$

and Logarithmic concave (L-concave) if

$$f(1 - \lambda)(x_1 + \lambda x_2) \geq (f(x_1))^{1-\lambda} \, (f(x_2))^{\lambda}.$$

**Definition 4.1.9**   A positive function $f$ defined on a convex set $K \subset R^n$ is said to be Harmonic convex (H-convex) if for each $x_1$, $x_2, \in K$ and $0 < \lambda < 1$,

$$f((1 - \lambda) x_1 + \lambda x_2) \leq \frac{1}{\dfrac{1 - \lambda}{f(x_1)} + \dfrac{\lambda}{f(x_2)}}$$

and Harmonic concave (H-concave) if

$$f((1 - \lambda)x_1 + \lambda x_1) \geq \frac{1}{\dfrac{1 - \lambda}{f(x_1)} + \dfrac{\lambda}{f(x_2)}} \, .$$

We shall now discuss some simple properties of L-convex and H-convex functions.

**Theorem 4.1.13**  Let $f$ be a positive function defined on a convex $K \subset R^n$.

   (i) If $f$ is H-convex on $K$, then it is L-convex and convex on $K$ but not conversely.

   (ii) If $f$ is concave on $K$, then it is L-concave and H-concave on $K$ but not conversely.

**Proof.**    (i) Let $f$ be H-convex on $K$. Then for $x_1, x_2 \in K$ and $0 < \lambda < 1$,

$$f((1 - \lambda)x_1 + \lambda x_2) \leq \frac{1}{\dfrac{1 - \lambda}{f(x_1)} + \dfrac{\lambda}{f(x_2)}}$$

$$\leq (f(x_1))^{1-\lambda} \, (f(x_2))^{\lambda}$$

$$\leq (1 - \lambda) f(x_1) + \lambda f(x_2).$$

Hence $f$ is L-convex and convex on $K$. To show that the converse is not true, let $f$ be any linear function on $R^n$ which is not a constant. Then for $f(x_1) \neq f(x_2)$ and $0 < \lambda < 1$,

$$f((1 - \lambda)x_1 + \lambda x_2) = (1 - \lambda) f(x_1) + \lambda f(x_2)$$

$$> \frac{1}{\dfrac{1 - \lambda}{f(x_1)} + \dfrac{\lambda}{f(x_2)}}$$

the strict inequality above follows from the fact that equality holds if and only if $f(x_1) = f(x_2)$.

(ii) Let $f$ be concave on $K$. Then for $x_1, x_2, \in K$ and $0 < \lambda < 1$,

$$f((1 - \lambda)x_1 + \lambda x_2) \geq (1 - \lambda) f(x_1) + \lambda f(x_2)$$

$$\geq (f(x_1))^{1-\lambda} \, (f(x_2))^{\lambda}$$

$$\geq \frac{1}{\dfrac{1 - \lambda}{f(x_1)} + \dfrac{\lambda}{f(x_2)}}$$

To show that the converse is not true, let $f(x_1) = (x)^2$ on

$$K = \{x \in R : x > 0\}.$$

It is clear that $f$ is not concave on $K$ (infact, it is strictly convex on $K$). However, it is H-concave on $K$ for if $x_1$, $x_2 \in K$ and $0 < \lambda < 1$ we have

$$f((1 - \lambda)x_1 + \lambda x_2) = ((1 - \lambda)x_1 + \lambda x_2)^2$$

$$\geq \frac{1}{\dfrac{1 - \lambda}{x_1} + \dfrac{\lambda}{x_2}}$$

$$\geq \frac{1}{\dfrac{1 - \lambda}{f(x_1)} + \dfrac{\lambda}{f(x_2)}}$$

$$= \frac{1}{\dfrac{1 - \lambda}{f(x_1)} + \dfrac{\lambda}{f(x_2)}} \; .$$

### Theorem 4.1.14

(i) Any positive linear function on a convex set $K \subset R^n$, except a positive constant, is not H-convex on $K$,

(ii) Each positive linear function on $K \subset R^n$ is H-concave on $K$.

**Proof.**  (i) Assume, to the contrary, that each positive linear function $f$ on $K \subset R^n$ is H-convex on $K$. Then for $x_1$, $x_2 \in K$ and $0 < \lambda < 1$,

$$f((1 - \lambda)x_1 + \lambda x_2) \leq \frac{1}{\dfrac{1 - \lambda}{f(x_1)} + \dfrac{\lambda}{f(x_2)}}$$

From the fact that $f$ is linear and arithmetic mean is greater than or equal to the harmonic mean we have

$$f((1 - \lambda)x_1 + \lambda x_2) = (1 - \lambda)f(x_1) + \lambda f(x_2)$$

$$\geq \frac{1}{\dfrac{1 - \lambda}{f(x_1)} + \dfrac{\lambda}{f(x_2)}}$$

Thus we have a contradiction and this proves (i). Proof of (ii) is easy and hence is omitted.

### Theorem 4.1.15  The reciprocal $f$ of a positive H-convex (concave) function $\varphi$ on $K \subset R^n$ is concave (convex) on $K$ and conversely.

**Proof.**  We have

$$\varphi((1-\lambda)x_1 + \lambda x_2) \geq \frac{1}{\dfrac{1-\lambda}{\varphi(x_1)} + \dfrac{\lambda}{\varphi(x_2)}}$$

Hence

$$f((1-\lambda)x_1 + \lambda x_2) \leq \frac{1-\lambda}{\varphi(x_1)} + \frac{\lambda}{\varphi(x_2)}$$

$$= (1-\lambda)f(x_1) + \lambda f(x_2).$$

This proves the result. The converse is trivial and hence is omitted.

**Theorem 4.1.16**  Let $f$ be H-concave on $K \subset R^n$. Then $f$ is strictly quasi-concave on $K$.

**Proof.**  Let $x_1, x_2 \in K$, $x_1 \neq x_2$, $f(x_2) > f(x_1)$ and $0 < \lambda < 1$. Then

$$f((1-\lambda)x_1 + \lambda x_2) \geq \frac{1}{\dfrac{1-\lambda}{f(x_1)} + \dfrac{\lambda}{f(x_2)}} > f(x_1)$$

and this completes the proof.

## 4.2  GENERALIZED CONVEX FUNCTION

We now discuss another concept related to convexity which was introduced by S. Nanda in the Journal of Indian Mathematical Society 72(2005).

We first quote the definition.

Let $\qquad f : R^n \to R$ and $g : R \to R$

$f$ is said to be g-convex on a convex set $K \subseteq R^n$ if the composite map $gf$ is convex on $K$; in other words, if for $\lambda \in [0, 1]$ and $x, y \in K$,

$$gf(\lambda x + (1-\lambda)y) \leq \lambda\, gf(x) + (1-\lambda)\, gf(y).$$

## Particular Cases

If $g$ is the identity map, then g-convexity of $f$ reduces to convexity, and if $g(x) = -x$, g-convexity implies that $f$ is concave. If $g$ is the log function and $f$ is g-convex, then

$$f(\lambda x + (1-\lambda)y) \leq (f(x))^{\lambda}(f(y))^{1-\lambda}$$

and this is the usual L-convexity of $f$ discussed by Klinger and Mangasanian in J. Math. Anal. Appl 24(1968) 388-408.

Observe that $g(x) = -\dfrac{1}{x}$ and $f$ is g-convex, if and only if $f$ is H-convex (harmonic convex) introduced C. Das which is given by

$$f(\lambda x + (1 - \lambda)y) \leq \frac{1}{\dfrac{\lambda}{f(x)} + \dfrac{1 - \lambda}{f(y)}}$$

for $x$, $y \in K$ and $\lambda \in [0, 1]$. Similarly $f(x) = \frac{1}{x}$ and $f$ is $g$-convex if and only if $f$ is H-concave.

### Theorem 4.2.1

(a) If $f$ is affine and $g$ is convex, then $f$ is $g$-convex.

(b) If $f$ and $g$ are convex and $g$ is nondecreasing, then $f$ is $g$-convex.

***Proof.***

(a) Since $f$ is affine, for $0 \leq \lambda \leq 1$ and $x$, $y \in K$,

$$f(\lambda x + (1 - \lambda)y) = \lambda f(x) + (1 - \lambda) f(y).$$

Since $g$ is convex,

$$gf(\lambda x + (1 - \lambda)y) = g(\lambda f(x) + (1 - \lambda) f(y))$$
$$\leq \lambda \, gf(x) + (1 - \lambda) \, gf(y)$$

and this proves (a).

(b) Since $f$ is convex, we have

$$f(\lambda x + (1 - \lambda)y) \leq \lambda f(x) + (1 - \lambda) f(y).$$

Since $g$ is nondecreasing and convex,

$$gf(\lambda x + (1 - \lambda)y) \leq g(\lambda f(x) + (1 - \lambda) f(y))$$
$$\leq \lambda \, gf(x) + (1 - \lambda) \, gf(y)$$

And this completes the proof.

### Theorem 4.2.2  $f$ is $g$-convex iff

$$G = \{(x, \alpha) : x \in K, \alpha \in R, gf(x) \leq \alpha\}$$

is a convex set.

***Proof.***  Let $G$ be convex and $x_1$, $x_2 \in K$. Then

$$(x_1, gf(x_1)) \text{ and } (x_2, gf(x_2)) \in G.$$

Since $G$ is convex,

$$(\lambda x_1 + (1 - \lambda)x_2, \lambda \, gf(x_1) + (1 - \lambda) \, gf(x_2) \in G$$

$$\Rightarrow gf(\lambda x_1 + (1 - \lambda)x_2) \le \lambda gf(x_1) + (1 - \lambda)\, gf(x_2).$$

$$\Rightarrow \qquad\qquad f \text{ is } g\text{-convex.}$$

Conversely, let $f$ be $g$-convex and $(x_1, \alpha_1)$, $(x_2, \alpha_2) \in G$. Then

$$gf(\lambda x_1 + (1 - \lambda)x_2) \le gf(x_1) + (1 - \lambda)\, gf(x_2)$$

$$\le \lambda \alpha_1 + (1 - \lambda)\alpha_2.$$

$$\Rightarrow \quad (\lambda x_1 + (1 - \lambda)x_2,\ \lambda \alpha_1 + (1 - \lambda)\alpha_2) \in G$$

$$\Rightarrow \qquad\qquad G \text{ is convex.}$$

**Theorem 4.2.3**  Let $f$ be $g$-convex.

Then the set

$$A_\alpha = \{x \in K : gf(x) \le \alpha\}$$

is convex for every real $\alpha$.

**Proof.**  Let $f$ be $g$-convex and let $x_1, x_2, \in A_\alpha$. Then

$$gf(\lambda x_1 + (1 - \lambda)x_2) \le \lambda gf(x_1) + (1 - \lambda)\, gf(x_2)$$

$$\le \lambda \alpha + (1 - \lambda)\alpha = \alpha.$$

$$\Rightarrow \qquad\qquad \lambda x_1 + (1 - \lambda)x_2 \in A_\alpha$$

$$\Rightarrow \qquad\qquad A_\alpha \text{ is convex.}$$

# Best Approximation

**5**

## 5.1 BEST APPROXIMATION

Let $X$ be a normed linear space. Suppose that an element $x$ in $X$ and a set $G$ in $X$ are given. Let $\delta$ be the distance from $x$ to the set $G$. By definition,

$$\delta = \inf_{g \in G} \|x - g\| \qquad \qquad \ldots(5.1)$$

The problem of best approximation consists in finding an element $g_0 \in G$ such that

$$\|x - g_0\| = \inf_{g \in G} \|x - g\| \qquad \qquad \ldots(5.2)$$

Every $g_0 \in G$ with this property is called an element of best approximation of $x$ (by elements of the set $G$). Denote by $f_G(x)$ the set of all elements of best approximation of $x$ by elements of the set $G$, i.e.

$$f_G(x) = \left\{ g_0 \in G : \|x - g_0\| = \inf_{g \in G} \|x - g\| \right\}$$

Observe that a best approximation $g_0$ is an element of minimum distance from the given $x$. Such a $g_0 \in G$ may or may not exist. This raises the problem of existence.

**Existence Theorem 5.1.1** (Existance) Let $G$ be a finite dimensional subspace of a normed linear space $X$. Let $x \in X$ be given. Then there exists an element of best approximation of $x$ by elements of $G$.

**Proof.** Consider the closed ball

$$S = \{g \in G : \|g\| \in 2\|x\|\}$$

Since $0 \in S$, it follows that

$$d(x, S) = \inf_{g_1 \in S} \|x - g_1\| \leq \|x - 0\|$$

$$= \|x\|$$

Now if $g \notin S$, then $\|g\| > 2\|x\|$

and $\qquad\qquad \|x - g\| \le \|g\| - \|x\| > \|x\| \le d(x, S)$ $\qquad\qquad$ ...(5.3)

This shows that $d(x, S) = d(x, G) = \delta$, and this value cannot be attained by a $g \in G - S$. Hence if a best approximation to $x$ exists, it must belong to $S$.

Note that instead of using the whole subset $G$ we have used only the closed ball $S$. The reason is the following: Since $S$ is closed and bounded and $G$ is finite dimensional, it follows that $S$ is compact. Hence, the norm is continuous and this implies that there is a $g_0 \in S$ such that $\|x - g\|$ assumes a minimum at $g = g_0$. By definition, $g_0$ is a best approximation to $x$.

**Examples 5.1.1**  (i) Consider the space $C\,[a, b]$. Let

$$G \;=\; \mathrm{span}\{x_0,\, x_1,\, ...,\, x_n\},\; x_j(t) = t^j$$

Note that $n$ is fixed here. This is the set of all polynomials of degree atmost $n$. it follows from Theorem (5.1.1) that for a given continuous function $x$ on $[a, b]$ there exists a polynomials $p_n$ of degree atmost $n$ such that fo every $g \in G$

$$\sup_{t \in [a,\, b]} \,|x(t) - P_n(t)| \;\le\; \left|\; \sup_{t \in [a,\, b]} \;\right| \,|x(t) - g(t)|$$

The approximation on $C\,[a, b]$ is called uniform approximation.

(ii) We shall see that finite dimensionality of $G$ in Theorem (5.1.1) is essential. Let $G$ be the set of all polynomials on $\left[0, \dfrac{1}{2}\right]$ of any degree. Consider $G$ as a subspace of $C\left[0, \dfrac{1}{2}\right]$. Then the dimension of $G = \infty$. Let

$$x\,(t) = (1 - t)^{-1}$$

Then for every $\epsilon > 0$ there is an $N$ such that $\|x - g_n\| < \epsilon$ for all $n > N$ where

$$g_n(t) = 1 + t + t^2 + ... + t^n$$

Hence $\qquad\qquad d(x, G) = 0$

However, since $x$ is not a polynomial, we find that there is no $g_0 \in G$ satisfying

$$\delta - d\,(x, G) = \|x - g_0\| = 0$$

We shall now see that for a given $x$ and $G$ there may be more than one best approximation.

The following is a simple example regarding the unique best approximation.

(iii) Let $X = R^3$ and $G$ be the $\zeta_1 \zeta_2$ – plane ($\zeta_3 = 0$). Then for a given point $x_0 = (\zeta_{10}, \zeta_{20}, \zeta_{30})$, a best approximation from the elements of $G$ is the point $g_0 = (\zeta_{10}, \zeta_{20}, 0)$. The distance from $x_0$ to $G$ is $\delta = |\zeta_{30}|$ and the best approximation $g_0$ is unique.

(iv) In this example we shall see that the best approximation is not unique. Let $X$ be the vector space of ordered pairs $x = (\zeta_1, \zeta_2) \ldots$ of real numbers with norm defined by

$$\|x\|_1 = |\zeta_1| + |\zeta_2| \qquad\qquad \ldots(5.4)$$

Take the point $x = (1, -1)$ and the subspace $G$ as as in (iii) above. Best approximation to $x$ from the elements of $G$ in the form defined by (5.4) is

$$G = \{y = (\eta, \eta) : \eta \text{ real}\}$$

It follows that

$$\|x - y\|_1 = |1 - \eta| + |-1 - \eta| \geq 2$$

The distance from $x$ to $G$ is $d(x, G) = 2$, and all $y = (\eta, \eta)$ with $|\eta| \leq 1$ are the best approximations to $x$ from the elements of $G$. This shows that for given $x$ and $G$ we have several best approximations, even infinitely many of them.

This raises the question of the uniqueness problem. We shall see that convexity will play a key role so far as this problem is concerned.

**Proposition 5.1.1**    (Convexity) Let $X$ be a normed linear space. In $X$ the set $E$ of best approximations to a given point $x \in X$ from the elements of a subspace $G$ of $X$ is convex.

***Proof.***    Let $\delta$ be the distance from $x$ to $G$. The theorem holds trivially if $E$ is empty or just one point. Suppose that $E$ has more than one point. Then for $y, z \in E$

$$\|x - y\| = \|x - z\| = \delta$$

We shall prove that this implies

$$w = \alpha y + (1 - \alpha) z \in E \ (0 \leq \alpha \leq 1) \qquad\qquad \ldots(5.5)$$

Since $w \in G$, $\|x - w\| \geq \delta$.

Also, $\|x - w\| = \| \alpha(x - y) + (1 - \alpha) (x - z)\|$

$$\leq \alpha \|x - y\| + (1 - \alpha) \|x - z\|$$

$$= \alpha\delta + (1 - \alpha) \delta = \delta$$

This shows that $\|x - w\| = \delta$. Hence $w \in E$. Since $y, z \in E$ are arbitrary, this proves that $E$ is convex.

*Remark 5.1.1*   If there are several best approximations to $x$ from the elements of $G$, then each of them lies in $G$. By definition, each of them has distance $\delta$ from $x$. It follows from Proposition (5.1.1) that $G$ and the closed ball

$$S = \{v : \|v - x\| \leq \delta\}$$

must have a segment $L$ in common. Hence, every $w \in L$ has distance $\|w - x\| = \delta$ from $x$. It follows that for each $w \in L$ there corresponds a unique

$$v = \frac{w - x}{\delta} \text{ of norm } \|v\| = \frac{\|w - x\|}{\delta} = 1$$

This shows that to each best approximation $w \in L$ given by (5.5) there corresponds a unique $v$ on $\{x : \|x\| = 1\}$. Thus, in order to obtain uniqueness of best approximations, we must exclude norms for which the set $\{x: \|x\| = 1\}$ can obtain segments of straight lines. The above remark suggests the following definition.

**Definition 5.1.1**   (Strict Convexity)   A normed linear space $X$ is called strictly convex if $\|x\| \leq r$ $\|y\| \leq r$ imply   $\|x + y\| < 2r$ unless $x = y$.

The strict convexity is also called rotund.

**Proposition 5.1.2** (a)  The Hilbert space is strictly convex.

(b)  The space $C[a, b]$ is not strictly convex.

**Proof.**   (a) Let $r > 0$. Write $\|x - y\| = r$ for all $x$ and $y$ $(x \neq y)$ of norm $1$. It follows from the parallelogram law that

$$\|x + y\|^2 + \|x - y\|^2 = 2(\|x\|^2 + \|y\|^2)$$

i.e. $$\|x + y\|^2 + r^2 = 2(1 + 1) = 4$$

i.e. $$\|x + y\|^2 < 4$$

i.e. $$\|x + y\| < 2$$

(b) Define $x_1$ and $x_2$ by

$$x_1(t) = 1, \ x_2(t) = \frac{t - a}{b - a}, \ t \in [a, b]$$

It follows that $x_1, x_2 \in C[a, b]$ and $x_1 \neq x_2$.

Also, $\|x_1\| = \|x_2\| = 1$, and

$$\|x_1 + x_2\| \ \sup_{t \in [a, b]} \left|1 + \frac{t - a}{t - b}\right| = 2$$

This shows that $C[a, b]$ is not strictly convex.

**Theorem 5.1.2** (uniqueness) Let $X$ be a normed linear space with a strictly convex norm. Then there is at most one best approximation to an $x \in X$ from the elements of a given subspace $G$ of $X$.

**Proof.**   Suppose there are two distinct best approximants to $x$, $u_1$ and $u_2$. Then $\|x - u_1\| = \|x - u_2\| = \delta$. Now $x - u_1$ and $x - u_2$ are also distinct. Hence by strict convexity,

$$\|(x - u_1) + ( x - u_2) \| < 2\delta$$

This is equivalent to $\|x - \frac{1}{2}(u_1 + u_2)\| < \delta$. But this would mean that the element $\frac{1}{2}(u_1 + u_2)$ is closer to $x$ than the minimum possible distance. This is a contradiction.

**Corollary 5.1.1**   The best approximation in the spaces $L_p[a, b]$, $1 < p < \infty$ is unique.

We shall now consider some problems in approximation theory which are directly related to optimization problems. We shall first formulate the problem.

Let $E$ be a subspace of an inner product space $X$. Let $x \in X$ be given. Find the element $z \in E$ close to $x$ in the sense that it minimizes $\|x - z\|$. Note that if $x$ lies in $E$, the solution is trivial. The following three important questions arise,

(i) Is there an element $z \in E$ which minimizes $\|x - z\|$?

(ii) Is the solution unique?

(iii) What is the solution or how is it characterized?

We shall briefly answer these questions.

**Theorem 5.1.3**   Let $E$ be a subspace of an inner product space $X$, and $x \in X$. If there is an element $y \in E$ such that $\|x - y\| \leq \| x - z\|$ for all $z \in E$, then $y$ is unique. A necessary and sufficient condition that $y \in E$ be a unique minimizing element in $E$ is that the error $x - y$ be orthogonal to $E$.

**Proof.**   We show first that if $y$ is a minimizing element, then $x - y$ is orthogonal to E.

Suppose there is an element $z \in E$ which is not orthogonal to $x - y$.

Assume that $\|z\| = 1$ and $< x - y, z > = \delta \neq 0$. Definey $y_1 = y + \delta z$. Then

$$\|x - y_1\|^2 = \|x - y - \delta z\|^2$$
$$= \|x - y\|^2 - < x - y, \delta z > - < \delta z, x - y > + |\delta|^2$$
$$= \|x - y\|^2 - |\delta|^2 < \|x - y\|^2$$

Thus, if $x - y$ is not orthogonal to $E$, $y$ is not a minimizing element.

We now show that if $x - y$ is orthogonal to $E$, then $y$ is a unique minimizing element.

It follows from Pythagorean theorem that for any $z \in E$,

$$\|x - z\|^2 = \|x - y + y - z\|^2 = \|x - y\|^2 + \|y - z\|^2$$

Hence,

$$\|x - z\| > \|x - y\| \text{ for } z \neq y$$

We shall now establish the existence of the minimizing element.

**Theorem 5.1.4**  Let $E$ be a non-empty closed subspace of a Hilbert space $H$. For every $x \in H$, there exists a unique $y \in E$ such that

$$\|x - y\| \leq \| x - z\|$$

for all $z \in E$. Furthermore, a necessary and sufficient condition that $y \in E$ be the unique minimizing element is that $x - y$ be orthogonal to $E$.

**Proof.**  If $x \in E$, then $y = x$ and the result holds triviality. Assume that $x \in E$ and define

$$\delta = \inf_{y \in E} \|x - z\| \qquad \qquad \ldots(5.6)$$

We have to show that there exists $y \in E$ with

$$\|x - y\| = \delta \qquad \qquad \ldots(5.7)$$

It follows from (5.6) that there is a sequence $Z_n$ in such that $\|x - Z_n\| \to$

$d$. For all $n$ and $m$, $\dfrac{(Z_n + Z_m)}{2} \in E$ so that $\left\| x - \dfrac{(Z_n + Z_m)}{2} \right\| \geq d$. Now by the parallelogram law,

$$\|Z_n - Z_m\|^2 + \|(Z_n + Z_m) - 2x\|^2$$

$$= 2\|Z_n - x\|^2 + 2\|Z_m - x\|^2$$

Hence,

$$\|Z_n - Z_m\|^2 \leq 2\|Z_n - x\|^2 + 2\|Z_m - x\|^2 - 4d^2$$

Since $\qquad \|Z_n - x\|, \|Z_m - x\| \to d \quad$ as $n, m \to \infty$

it follows that

$$\|Z_n - Z_m\|^2 \to 0 \qquad \qquad \text{as } m, n \to \infty$$

This shows that $Z_n$ is a Cauchy sequence, since $E$ is a closed subspace of the Hilbert space $H$, the sequence $Z_n$ converges to some $y \in E$. It follows from the continuity of the norm that $\|x - y\| = \lim \|x - Z_n\| = d$. Thus the existence of the minimizing element is established.

The uniqueness and orthogonality have been established in Theorem (5.1.3).

Observe that according to Theorem (5.1.4) there exists a unique best approximation $y$ to $x$ from $E$ and this $y$ is the unique element for which $x - y$ is orthogonal to $E$.

**Minimum distance to a convex set**   In the previous section we have discussed the minimum norm problem from a given element to a linear subspace. We shall now extend Theorem (5.1.4) to more general subsets. The following theorem provides a sufficient condition which is useful from the application point of view.

**Theorem 5.1.5**   Let $E$ be a non-empty closed, convex subset of Hilbert space $H$. Let $x \in H$ and let

$$d = \inf_{y \in E} \|x - y\| \qquad \qquad \ldots(5.8)$$

Then there is a unique $y_0$ in $E$ such that

$$\|x - y_0\| = d \qquad \qquad \ldots(5.9)$$

**Proof.**   It follows from (5.8) that

$$\lim_{n \to \infty} \|x - y_n\| = d \qquad \qquad \ldots(5.10)$$

By parallelogram law

$$\|y_m - y_n\|^2 = 2\|y_m - x\|^2 + \|y_n - x\|^2 - \|2x - y_m - y_n\|^2$$

$$= 2\|y_m - x\|^2 + 2\|y_n - x\|^2 - 4\left\|x - \frac{1}{2}(y_m + y_n)\right\|^2$$

Since $E$ is convex, it follows that

$$\frac{1}{2}(y_m + y_n) \in E \text{ and } \left\|x - \frac{1}{2}(y_m + y_n)\right\|^2 \geq d^2$$

Therefore,

$$\|y_m - y_n\|^2 \leq 2\|x - y_m\|^2 + 2\|x - yn\|^2 - 4d^2$$

Using (5.10) we see that

$$\|y_m - y_n\| \to \infty \text{ as } m, n \to \infty$$

This shows that $\{y_n\}$ is a Cauchy sequence. Since $H$ is complete, there is a $y_0 \in H$ such that $\|y_m - y_n\| \to 0$ Since $E$ is closed, $y_0 \in H$. Now

$$\|x - y_0\| \le \| x - y_n\| + \|y_n - y_0\| \to d + 0 = d$$

By definition,

$$\|x - y_0\| \ge d$$

Hence

$$\|x - y_0\| = d$$

We now show that $y_0$ is unique. Suppose $y_0$ and $y_1$ are two elements

where
$$\|x - y_0\| = \|x - y_1\| = d.$$

Since $E$ is convex, $\frac{1}{2}(y_0 + y_1) \in E$.

Hence,

$$d \le \left\| x - \frac{1}{2}(y_0 + y_1) \right\| = \left\| \frac{1}{2} x - \frac{1}{2} ya + \frac{1}{2} x - \frac{1}{2} y_1 \right\|$$

$$\le \frac{1}{2}\|x - y_0\| + \frac{1}{2}\|x - y_1\| = \frac{d}{2} + \frac{d}{2} = d$$

Thus

$$\left\| x - \frac{1}{2}(y_0 + y_1) \right\| = d$$

It now follows from Parallelogram law, that

$$\|y_0 - y_1\|^2 = 2\|x - y_0\|^2 + 2\|x - y_1\|^2 - 4 \left\| x - \frac{1}{2}(y_0 + y_1) \right\|^2$$

$$= 2d^2 + 2d^2 - 4d^2 = 0$$

Hence $y_0 = y_1$.

We shall now give an application of Theorem (5.1.5) which is directly related to problems on approximation theory.

**Theorem 5.1.6**    Let $E$ be the set of polynomials of degree $\le n$ that are convex on $[a, b]$. Let $f \in L^2 [a, b]$. Then the problem

$$\min_{p \in E} \|f - p\|, \ \|f\|^2 = \int_a^b |f|^2 \, dx \tag{5.11}$$

has a unique solution.

**Proof.**    Note that a polynomial $p(x)$ is convex on $[a, b]$ only if $p''(x) \ge 0$ on $[a, b]$. Also if $p$ and $q$ are convex on $[a, b]$ then

$$\alpha p(x) + (1 - \alpha) \, q(x)$$

is convex on $[a, b]$ for $0 \le \alpha \le 1$. It follows that $\alpha p'' + (1 - \alpha) \, q'' \ge 0$. Hence, $E$ is convex. We now show that $E$ is a closed subset of $L^2 [a, b]$. Let $p_k(x) \in E$

Suppose that $p_k(x)$ cnverges to $f(x) \in L^2 [a, b]$, i.e.

$$\lim_{k \to \infty} \int_a^b |f(x) - p_k(x)|^2 \, dx = 0$$

Let $q(x)$ be the best approximation to $f(x)$.
Then

$$0 \le \int_a^b |f(x) - q(x)|^2 \, dx \le \int_a^b |f(x) - p_k(x)|^2 \, dx$$

Letting $k \to \infty$ we see that

$$\int_a^b |f(x) - q(x)|^2 \, dx = 0$$

Hence $f(x) = q(x)$ is in $P_n$.

Let $P_0^*(x)$, $P_1^*(x)$, ..., $P_n^*(x)$ denote the orthogonal polynomials.
Then

$$f(x) - p_k(x) = a_{0k} P_0^*(x) + \ldots + a_{nk} P_n^*(x), \; k = 1, 2, 3, \ldots$$

for some constants $a_{tk}$. It follows that

$$\int_a^b |f(x) - p_k(x)|^2 \, dx = \sum_{i=0}^{n} |a_{jk}|^2 \to 0$$

Thus, $\lim_{k \to \infty} a_{ik} = 0, \quad i = 1, 2, \ldots, n$ \hfill (5.12)

In view of (5.12), for any bounded region $\Omega$ in the complex plane

$$|f(z) - p_k(z)| \le \sum_{i=1}^{n} |a_{jk}| \, |P_i^*(z)| \text{ as } k \to \infty$$

i.e.,

$$p_k(z) \to f(z) \qquad \text{uniformly in } \Omega$$

$$\text{Hence, } p''k(z) \to f''(z) \qquad \text{uniformly in } \Omega.$$

$P''_k(x) \ge 0$ on $[a, b]$, it follows that $f''(x) \ge 0$ on [a, b]. Thus $E$ is closed. The theorem now follows by an application of Theorem (5.1.5).


## 5.2   ROTUND

In this section we first discuss an equivalent characterization of rotund.

Recall that a duality map $J : X \to 2^{x^*}$ satisfies

$$J(x) = \{f \in X : f(x) = \|f\| \, \|x\|, \, \|f\| = \|x\| \}.$$

We say that $J$ is strictly monotone if for every $x \ne y$ and every $f \in J(x)$, $g \in J(y)$

We have

$$(f - g)(x - y) > 0$$

For $\quad f \in J(x),\ g \in J(y) \quad$ we have

$$(f - g)(x - y) = (\|x\| - \|y\|)^2 + (\|f\|\,\|y\| - f(y)) + (\|g\|\,\|x\| - g(x))$$

We have the following characterization.

**Theorem 5.2.1**  $X$ is strictly convex iff $J$ is strictly monotype.

**Proof.**  Since each term of Browder's expansion is nonnegative, $J$ is not strictly monotone

iff $\qquad\qquad \exists x \neq y,\ f \in J(x),\ g \in J(y)$

with $\qquad\qquad (f - g)(x - y) = 0$

iff $\qquad\qquad \exists f \in X^*$ with

$$f\!\left(\frac{\alpha}{\|x\|}\right) = f\!\left(\frac{y}{\|y\|}\right) = \|f\| = \|x\| = \|y\|$$

iff $\exists f \in X^*$ that attains a maximum at two points of the unit sphere.

This proves the result.

There are several other properties, which are equivalent to Rotund $(R)$. In order to mention a few of them we need some definitions.

**Definition 5.2.1**  (a) A wedge $W$ in a linear space $L$ is a convex set $W$ such that $tw \subseteq W$ for every $t \geq 0$.

(b) A cone is a wedge which contains no line through the origin.

(c) The half space $\{x : f(x) \leq 0\}$ is a wedge but is not a cone.

(d) A wedge $W$ is a cone iff $x \in W,\ -x \in W \Rightarrow x = 0$.

(e) Let $K$ be a convex set in a linear space $L$. A point $x \in K$ is called a passing point of $K$ if $x$ belongs to an open segment which is contained in $K$.

(f) $x \in K$ is an extreme point of $K$ iff

(i) $x = \dfrac{x_1 + x_2}{2}$ and $x_i \in K$ imply that $x = x_1 = x_2$, and if and only if

(ii) Whenever $x$ is a convex combination of a finite set of points of $K$, then $x$ is an element of $\varphi$.

Let $X$ be a Banach space. Then

$$U = \{x \in X : \|x\| \leq 1\}$$
$$S = \{x \in X : \|x\| = 1\}$$
$$U^* = \{f \in X^* : \|f\| \leq 1\}$$
$$S^* = \{f \in X^* : \|f\| = 1\}$$

We now state some equivalent formulations of Rotund $(R)$ and we omit the proof.

$R_1$.　Every $f \in X^*$ attains a maximum on atmost one point of $S^*$.

$R_2$.　Every point of $S$ is an extreme point of $U$.

$R_3$.　Every point of $S$ is an exposed point of $U$.

$R_4$.　Every hyperplane of support of $U$ touches $U$ in atmost one point.

$R_5$.　Supporting hyperplanes to $U$ at distinct point of $S$ are distinct.

$R_6$.　If $W$ is a wedge in $X$ in which norm is additive, then $W$ is a half ray $\{tx : t \geq 0\}$.

$R_7$.　In every two dimensional linear subspace $P$ of $X$, unit ball $P \cap U$ is rotund.

We now discuss some other concepts like Uniform convexity and Local uniform convexity in a nls.

**Definition 5.2.2**　$X$ is said to be uniformly convex (uniformly Rotund or $UR$ in short) if for all $\epsilon > 0$, $x, y \in U$ and $\|x - y\|$ imply that $\left\|\frac{1}{2}(x + y)\right\| \leq 1 - \delta$

where $\delta = \delta(\epsilon)$ is independent of $x$ and $y$ and $0 < \delta < 1$. A reformulation of $UR$ is as follows: If $x_n \in U$, $y_n \in U$, $\|x_n + y_n\| \to 2$, then $\|x_n - y_n\| \to 0$.

An application of parallelogram law proves the following.

**Theorem 5.2.2**　Every ips $X$ is $UR$.

**Proof.**　Let $x, y \in U$ and $\|x - y\| > \epsilon$.

By parallelogram law we have

$$\left\|\frac{x + y}{2}\right\|^2 = \frac{1}{2}\|x\|^2 + \frac{1}{2}\|y\|^2$$

$$- \frac{1}{4}\|x - y\|^2$$

$$< \frac{1}{2} + \frac{1}{2} - \frac{1}{4}\epsilon^2$$

$$= 1 - \left(\frac{\epsilon}{2}\right)^2$$

Therefore we can find a number $\delta > 0$ such that $\left\|\dfrac{x + y}{2}\right\| \leq 1 - \delta$.

**Definition (LUR) 5.2.3**  $X$ is said to be locally uniformly convex (rotund) and *LUR* in short if for each $\epsilon$ with $0 < \epsilon \leq 2$ and each $x \in S$, there is a $\delta = \delta\,(\epsilon, x) > 0$ such that if $y \in U$ and $\|x - y\| \geq \epsilon$, then $\|x + y\| \leq 2(1 - \delta\,(\epsilon, x))$.

Equivalently

**(LUR)'** $$x \in S,\ x_n \in U,\ \|x_n + x\| \to 2$$
$$\Rightarrow \|x_n + x\| \to 0.$$

We shall now quote the definition of a generalization of *LUR* and hence of *UR*. We call this concept *(CLUR)*.

**(CLUR)** $$x \in S,\ x_n \in U,\ \|x_n + x\| \to 2$$

$\Rightarrow$ the sequence $(x_n)$ is compact in $U$, i.e. $(x_n)$ has a convergent subsequence in $U$.

**Example 5.2.1**  A reflexive space which is (CLUR) but not (LUR).

Let $$X = l_p\ (1 < p < \infty).\text{ Let } x \in X.$$
Denote,
$$x' = (0, x_2, x_3, \ldots, x_n, \ldots)$$
$$x'' = (x_1, 0, x_3, \ldots, x_n, \ldots)$$

Define a new norm in $X$ by

$$|||x||| = \max\left\{ \|x'\|_p,\ \|x''\|_p \right\}.$$

Clearly $$\frac{1}{2}\|x\|_p \leq |||x||| \leq \|x\|_p,$$

showing that the two norms are equivalent. It is easy to see that $|||\ |||$ is not $(R)$ and hence cannot be (LUR). To prove that it is (CLUR) we take $x, x_n \in X$ such that $|||x||| = 1,\ |||x_n||| \leq 1,\ |||x + x_n||| \to 2$.

There are two possibilities:

(i) $\|x' + x_n'\|_p \geq \|x'' + x_n''\|_p$ for infinity of $n$.

(ii) $\|x' + x_n''\| < \|x'' + x_n''\|_p$

Suppose (i) holds. Since we are interested in extracting a convergent subsequence of $(x_n)$ we may assume without loss of generality that (i) holds for all sufficiently large $n$. In this case $\|x' + x_n'\|_p \to 2$, which shows that $\|x'\|_p = 1$. $\|x_n'\|_p \to 1$ and then by the (LUR) property of the $l_p$ – norm $\|x_n' - x'\| \to 0$. As the sequence of real numbers $(x_n(1))$ is bounded, it follows that the sequence $(x_n)$ is compact.

Case (ii) is similarly proved.

**Example 5.2.2**  Example of a nonreflexive space which is (CLUR) but not (LUR).

Let $X = c_0$ and $x \in X$. Let $x'$, $x''$ be denoted as in the previous example.

Let $\| \ \|_1$ be the norm constructed by M.M. Day. It is known that this norm is LUR.

Now we define a new norm in $c_0$ by

$$\||x\|| = \max\{ \|x'\|_1, \|x''\|_1 \}.$$

It can be shown by computation that

$$\frac{1}{2}\|x\|_1 \leq \||x\|| \leq \|x\|_1,$$

and hence the two norms are equivalent. The proof that it is (CLUR) is similar to that of the previous example.

**Theorem 5.2.2**  (M. Edelstein) Let $X$ be a separable conjugate Banach space and let $S$ be a closed, bounded set in $X$. Then there is an element $x \in X$ and $s \in S$ such that

$$\|x - s\| = \inf \{ \|x - y\| : y \in S \}.$$

*Remark 5.2.1*  It may be noted that $X$ must be both separable and conjugate in order that the result holds.

For example if $X = c_0$, which is separable but not conjugate, than the theorem fails to hold. In order to show this, let

$$A_n = \{x = (x_k) \in c_0 : |x_k| = 1 - \frac{1}{n+1} \text{ for } k \leq n, \ x_k = 0 \text{ for } k > n\}$$

It is easy to show that $\overset{\infty}{\underset{n=1}{\cup}} A_n$ is closed and bounded. If $y \in c_0$ and $\|y\| \geq 3$,

then $y$ has no best approximation in $A = \overset{\infty}{\underset{n=1}{\cup}} A_n$. For suppose $x \in A_n \subset A$ is

nearest to $y$. Then clearly $\|x - y\| \geq 2$. Let $m > n$ be such that

$$|y_k| < \frac{1}{2}\|x - y\| \text{ for } k > m \text{ and}$$

choose $\bar{x} = (\bar{x}_k) \in A_m$ so that

$$\frac{\bar{x}_k}{|\bar{x}_k|} = \frac{y_k}{|y_k|} \text{ whenever } x_k \text{ and } y_k \text{ are both nonzero. Then}$$

$$|y - \bar{x}_k| = |y_k| < \frac{1}{2} \|x - y\| \text{ for } k > m;$$

$$|y - \bar{x}_k| \le |y_k| + |\bar{x}_k| < 2 \le \|x - y\|$$

for
$$|y_k| \le 1,$$

and
$$|y_k - \bar{x}_k| = |y_k| < \left| \frac{|y_k|}{|x_k|} \bar{x}_k - \bar{x}_k \right|$$

$$= |\bar{x}_k| \left| \frac{|y_k|}{\bar{x}_k} - 1 \right|$$

$$= |y_k| - 1 + \frac{1}{m + 1}$$

$$< |y_k| - 1 + \frac{1}{n + 1}$$

$$\le \|x - y\|.$$

Thus
$$\|\bar{x} - y\| < \|x - y\|$$

contrary to the assumption that $x$ is nearest to $y$.

## 5.3  FARTHEST POINTS

Let $X$ be a nls and $S$ a bounded subset of $X$. Define a map

$$r : X \to R \text{ by}$$

$$r(x) = \sup\{ \|x - s\| : s \in S \}.$$

Observe that this mapping $r$ is convex; it is in fact the supremum of convex functions. It is continuous as well, in fact

$$|r(x) - r(y)| \le \|x - y\|.$$

A point $s_0 \in S$ is called a farthest point in $S$ (from the elements of $X$) if there exists a point $x \in X$ such that

$$\|x - s_0\| = r(x) = \sup_{s \in S} \|x - s\|$$

Let $b(s)$ denote the set of all farthest points in $S$ from the elements of $X$. Let $a(S)$ denote the set of all points in $X$ for which there is a farthest point $s \in S$.

We now state (without proof) the following existence theorem for farthest points.

**Theorem 5.3.1** (Panda and Kapoor) Let $S$ be a bounded closed subset of a reflexive (CLUR) space $X$, then $b\,(S)$, the set of all farthest points in $S$ from the elements of $X$ is nonempty. Further the set $a\,(S)$ is dense in $X$.

## Best Simultaneous Approximation

Let $C$ be a subset of a normed linear space $X$. Given any bounded subset $F$ in $X$, define

$$d(F, c) = \inf_{\alpha \in C} \sup_{f \in F} \|f - c\|$$

An element $c_0 \in C$ is said to be best simultaneous approximations (bsa) to $F$ if

$$d(F, c) = \sup_{f \in F} \|f - c_0\|$$

An element $c_0 \in C$ is said to be a simultaneous farthest point (sfp) if

$$\sup_{c \in C} \sup_{f \in F} \|f - c\| = \sup_{f \in F} \|f - c_0\|$$

**Lemma 5.3.1** Let $c \in X$ and $F$ a bounded subset of $X$. Then

$$\varphi(c) = \sup_{f \in F} \|f - c\|$$

is a continuous functional on $X$.

**Proof.** For any $f \in F$ and $c, c' \in X$,

We can write

$$\|f - c\| \leq \|f - c'\| + \|c - c'\|$$

Then

$$\sup_{f \in F} \|f - c\| \leq \sup_{f \in F} \|f - c'\| + \|c - c'\|$$

Now if $\|c - c'\| < \epsilon$, then

$$\varphi(c') \leq \varphi(c') + \epsilon$$

On interchanging $c$ and $c'$ we obtain

$$\varphi(c') \leq \varphi(c) + \epsilon$$

and this completes the proof.

**Lemma 5.3.2** Let $C$ be a convex subset of $X$ and $F \subset X$. If $c_1, c_2 \in C$ are bsa to $F$ by  elements of $C$ then

$$c = \lambda c_1 + (1 - \lambda)\, c_2, \qquad 0 \le \lambda \le 1 \text{ is a bsa to } F.$$

**Proof.**  We have

$$\sup_{f \in F} \|f - c\| = \sup_{f \in F} \| \lambda(f - c_1) + (1 - \lambda)\,(f - c_2)\|$$

$$\le \sup_{f \in F} \|f - c_1\| + (1 - \lambda)\sup_{f \in F} \|f - c_2\|$$

$$= \lambda d(F, c) + (1 - \lambda)\, d(F, c)$$

$$= d(F, c).$$

The reverse inequality is always true and this completes the proof.

**Theorem 5.3.2**  Let $X$ be strictly convex and $C$ a weakly compact convex subset of $X$. Then there is a unique bsa from the elements of $c$ to any given compact subset $F$ of $X$.

**Proof.**  Since $\varphi$ defined in Lemma 5.3.1 is convex and continuous, it is a weakly lower semicontinuous function on $c$. Since $c$ is weakly compact, $\varphi$ attains its infimum at $c_0 \in C$ say. Therefore

$$d(F, c) = \sup_{f \in F} \|f - c_0\|.$$

Now for the uniqueness, let

$c_1, c_2, c_1 \ne c_2$ be two best simultaneous approximations by elements of $F$, i.e.

$$\sup_{f \in F} \|f - c_1\| = \sup_{f \in F} \|f - c_2\| = d \text{ say}$$

Then

$$\sup_{f \in F} \left\| f - \frac{c_1 + c_2}{2} \right\| = d.$$

Since $F$ is compact, there exists a $f_0 \in F$ with

$$\sup \left\| f - \frac{c_1 + c_2}{2} \right\| = \left\| f_0 - \frac{c_1 + c_2}{2} \right\| = d,$$

it than follows that

$$\|f_0 = c_1\| \text{ and } \|f_0 - c_2\|$$

Since $X$ is strictly convex we get $c_1 = c_2$

**Theorem 5.3.3**  Let $X$ be strictly convex and $C$ a reflexive subspace of $X$. Then for any nonempty compact subset $F$ of $X$ there exists one and only one bsa in $C$.

**Proof.**   Since $F$ is compact, $\|f\| \leq M$ for every $f \in F$. Take the ball

$$B = B\,(0,\,2M) \subset C. \text{ Then}$$

$$\inf_{c \in B}\ \sup_{f \in F}\|f - c\| = \inf_{c \in C}\ \sup_{f \in F}\|f - c\| \leq M.$$

The ball $B$ is weakly compact in $C$ and $\varphi$ is a weakly lower semicontinous function on $B$. Therefore $\varphi$ attains its infimum in $B$ for some $c_0 \in B$, say, which is a bsa to $F$, i.e.

$$d(F,\,c) = \sup_{f \in F}\|f - c_0\|.$$

Hence uniqueness of bsa follows.

# Fixed Point Theory

**6**

We shall first discuss the famous Banach fixed-point theorem (also called the Banach contraction mapping principle). There are several far-reaching consequences of this theorem and we shall discuss some of these. First of all we have some definitions.

**Definition 6.1.1**  Let $(X, d)$ be a metric space and let $f : (X, d) \to (X, d)$, $f$ is called a contraction if there is a real $\alpha < 1$ such that

$$d(f(x), f(y)) < ad(x,y)$$

for all points $x, y \in X$.

**Definition 6.1.2**  Let $(X, d)$ be a metric space. A point $x \in X$, if it exists, is called a fixed point of a mapping

$$f : (X, d) \to (X, d)$$

if $f(x) = x$.

## 6.1  BANACH FIXED POINT THEORY

Let $T$ be a contraction mapping on a complete metric space $X$. Then $T$ has a unique fixed point.

**Proof.**  Let $x_0$ be an arbitrary point in $X$. Define the 'iterative sequence' $(x_n)$ by

$$x_0, \; x_1 = Tx_0, \; x_2 = Tx_1 = T^2 x_0, \; ..., \; x_n = T^n x_0, \qquad ...(6.1)$$

If $m > n$, then

$$
\begin{aligned}
d(x_m, x_n) &= d(T^m x_0, T^n x_0) \\
&= d(T^m X_0, T^m T^{n-m} x_0) \\
&\leq a^m d(x_0, T^{n-m} x_0) \\
&= \alpha^m d(x_0, x_{n-m}) \\
&\leq \alpha^m (d(x_0, x_1) + ... + d(x_{n-m-1}, x_{n-m})) \\
&\leq \alpha^m d(x_0, x_1) (1 + \alpha + ... + \alpha^{n-m-1})
\end{aligned}
$$

$$\leq \frac{\alpha^m(d(x_0, x_1))}{1 - \alpha} \qquad \ldots(6.2)$$

Since $\alpha < 1$ and $d(x_0, x_1)$ is fixed, we can make the right side of (6.2) as small as we please by taking m sufficiently large. This shows that $(x_n)$ is a Cauchy sequence. Since $X$ is complete, there exists a point $x$ in $X$ such that $x_n \to x$.

We now show that $x$ is a fixed point of the mapping $T$. It follows from the triangle inequality and (6.1) that

$$d(x, Tx) \leq d(x, x_n) + d(x_n, Tx)$$

$$\leq d(x, x_n) + \alpha d(x_{n-1}, x).$$

Since $x_n \to x$, it follows that $d(x, Tx) = 0$ and hence $Tx = x$. This shows that $x$ is a fixed point of $T$.

We conclude the proof by showing that $x$ is unique. Suppose that $x'$ is also a fixed point; that is, $Tx' = x'$; then

$$d(x, x') = d(Tx, Tx') \leq \alpha d(x, x')$$

and since $\alpha < 1$, this implies that

$$d(x, x') = 0 \text{ and hence } x = x'.$$

### 6.1.1   An application

We shall now give an application of the Banach fixed point theorem to linear equations. Let $X = R^n$. Define a metric $d$ on $R^n$ by

$$d(x, z) = \max |x_i - z_i| \qquad \ldots(6.3)$$

where $x = (x_1, \ldots x_n)$ and $z = (z_1, \ldots z_n)$. Note that $X$ with this metric is complete. Define now a map $T : X \to X$ by

$$y = Tx = Ax + b \qquad \ldots(6.4)$$

where $A = (\alpha_{jk})$ is a fixed real $n \times n$ matrix and $b \in X$ a fixed vector. Write the equation (6.4) in the following form:

$$y_j = \sum_{k=1}^{n} \alpha_{jk} x_k + \beta_j \, j = 1, 2, \ldots, n,$$

where $b = (\beta_j)$. Set $\omega = (\omega_j) = Tz$. Using the equations (6.3) and (6.4) we find that

$$d(y, \omega) = d(Tx, Tz) = \max_j |y_j - \omega_j|$$

$$= \max_j \sum_{k=1}^{n} \alpha_{jk}(x_k - \omega_j)$$

$$\leq \max_{i} |x_i - z_i| \max_{j} \sum_{k=1}^{n} |\alpha_{jk}|$$

$$= d(x, z) \max_{j} \sum_{k=1}^{n} |\alpha_{jk}|t$$

$$= \alpha d(x, z)$$

where $\alpha = \max_{j} \sum_{k=1}^{n} |\alpha_{jk}|$. Thus $T$ is a contraction if $x < 1$. Therefore we have the following theorem.

**Theorem 6.1.1**    Let $A = (\alpha_{jk})$ be $n \times n$ matrix. Let $x = Ax + b$ be a system of n linear equations in $n$ variables where

$$\sum_{k=1}^{n} |\alpha_{jk}| < 1 \quad j = 1, 2, ..., n.$$

Then the system has precisely one solution. This solution is given by the limit of the iterative sequence

$$X^{(n+1)} = Ax^{(n)} + b \quad n = 0, 1, 2, ...$$

where $x^{(0)}$ is arbitrary.

We shall now use the Banach contraction mapping principle to prove the famous Picard's theorem.

**Picard's Theorem  6.1.2**    Let $f(x, y)$ and $\dfrac{\partial f}{\partial y}$ be continuous on a closed rectangle $E = \{(x, y) : a_1 \leq x \leq a_2, b_1 \leq y \leq b_2\}$, and let $(x_0, y_0)$ be an interior point of $E$. Then the differential equation

$$\frac{dy}{dx} = f(x, y) \qquad\qquad ...(6.5)$$

has a unique solution $y = g(x)$ which passes through $(x_0, y_0)$.

**Proof.**    Since $f(x, y)$ and $\dfrac{\partial f}{\partial y}$ are continuous on $E$, they are bounded. Hence there exist constants $K$ and $M$ such that

$$|f(x, y)| \leq K$$

and $\left| \dfrac{\partial}{\partial y} f(x, y) \right| \leq M$

for all points $(x, y)$ in $E$. Let $(x, y_1)$ and $(x, y_2)$ belong to $E$. It follows from the mean value theorem that

$$|f(x, y_1) - (x, y_2)| = |y_1 - y_2| \left| \frac{\partial}{\partial y} f(x, y_1 + \theta(y_2 - y_1)) \right|$$

for some $\theta$, $0 < \theta < 1$. Thus

$$|f(x, y_1) - f(x, y_2)| \le M|y_1 - y_2|$$

for all $(x, y_1)$ and $(x, y_2)$ in $E$. Note that the only use we make of the assumption that $\dfrac{\partial f}{\partial y}$ exists and is continuous on $E$ is to derive the Lipschitz condition stated above.

We shall now replace our problem by an equivalent problem through an integral equation. Let $y = g(x)$ be such that

$$\frac{dy}{dx} = f(x, y) \text{ where } y_0 = g(x_0).$$

Then integrating the expression from $x_0$ to $x$ we have

$$g(x) - g(x_0) = \int_{x_0}^{x} f(u, g(u))\, du, \text{ that is,}$$

$$g(x) = y_0 + \int_{x_0}^{x} f(u, g(u))\, du \qquad \qquad \text{...(6.6)}$$

Conversely, if $y = g(x)$ satisfies the equation (6.6) then $y_0 = g(x_0)$. Differentiating the equation (6.6) we get (6.5). Thus it is sufficient to prove that the integral equation (6.6) has a unique solution.

Choose a positive number $c$ such that $Mc < 1$. Consider the closed rectangle $F$ determined by $|x - x_0| \le c$ and $|y - y_0| \le Kc$ which is contained in $E$. Let $G$ be the set of all continuous real functions $y = g(x)$ defined on $|x - x_0| \le c$ such that $|g(x) - y_0| \le Kc$. Note that $G$ is a closed subspace of the complete metric space $C[x_0 - c, x_0 + c]$ and is therefore itself a complete metric space.

Consider now the mapping $T$ of $G$ into itself defined by $Tg = h$, where

$$h(x) = y_0 + \int_{x_0}^{x} f(u, g(u))\, du$$

Since $f$ is bounded, it follows that

$$|h(x) - y_0| = \left| \int_{x_0}^{x} f(u, g(u))\, du \right| \le Kc$$

Furthermore, since $\dfrac{\partial}{\partial y} f$ is bounded, it follows that

$$|h_1(x) - h_2(x)| = \left| \int_{x_0}^{x} \{f(u, g_1(u)) - f(u, g_2(u))\}\, du \right|$$

$$\le M\, c \, \sup |g_1(x)\, g_2(x)|$$

Since $Mc < 1$, this shows that $T$ is a contraction mapping of $C$ into itself. Applying the Banch fixed point theorem we conclude that the equation $Tg = g$ has a unique solution. This completes the proof of Picard's theorem.

## 6.2 BROUWER'S FIXED POINT THEOREM AND ITS APPLICATION

The most important result in the fixed-point theory is famous theorem of Brouwer.

**Brouwer's Theorem 6.2.1**  Every continuous map of the closed unit ball $S = \{x : \|x\| \le 1\}$ in $R^n$ to itself has a fixed point.

There are several proofs of this classic result, but since they all depend on the methods of algebraic topology, we omit the proof.

In order to give an important application of Brouwer's theorem we need an equivalent form of this theorem.

**Theorem 6.2.2**  Every continuous map of a closed bounded convex set in $R^n$ into itself has a fixed point.

In order to demonstrate the application of Brouwer's theorem we need an example of continuous operators in $R^n$ given by a matrix.

**Example 6.2.1**  Let $C^n$ be the space of $n$-tuples of complex numbers. Let $A$ be an $m \times n$ matrix. Define $a$ map $T : C^n \to C^m$ by

$$Tx = Ax \text{ for } x \in C^n$$

It follows from the Cauchy-Schwarz inequality that for any $i$.

$$\left| \sum_j A_{ij}(x_j - y_j) \right| \le \left( \sum_j |A_{ij}|^2 \right)^{\frac{1}{2}} \|x - y\|$$

Hence,

$$\|Ax - Ay\| = \|A(x - y)\|$$

$$\le \left( \sum_{i,j} |A_{ij}|^2 \right)^{1/2} \|x - y\|$$

$$\le \sqrt{nm}\, M \|x - y\|$$

where $M$ is the modulus of the largest element of $A$. Hence, given $\varepsilon > 0$ the continuity condition is satisfied by taking $\delta = \varepsilon/M \sqrt{nm}$ for any $x$. Thus the operator is uniformly continuous.

The following result is an application of Brouwer's theorem to the theory of matrices, which is important in many applications.

**Perron's Theorem 6.2.3**   Let A be a matrix whose elements are all positive. Then A has at least one positive eigenvalue and the elements of the corresponding eigenvectors are all non-negative.

**Proof.**   Note that a positive matrix transforms vectors with positive components into vectors with positive components. Geometrically this means working in the n-dimensional analogue of the first quadrant. The eigenvalue equation concerns vectors mapped into multiplies of themselves. Thus, it is reasonable, when looking for eigenvectors, to consider only vectors with positive components. If we could somehow disregard the magnitude of the vectors and concentrate on its direction, then an eigenvector could be regarded as a "fixed" vector under the transformation, in the sense that its direction would be unchanged. We therefore, consider $\dfrac{x}{\|x\|}$ in place of $x$. We would then apply the operator to the n-dimensional analogue of the positive quadrant of the surface of the unit sphere. The surface of a sphere, however, is not convex. We shall therefore use a plane rather than a surface of the sphere. Define

$$E = \left\{ x \in R^n : x_1 \geq 0 \text{ for each } i \text{ and } \sum_{i}^{n} x_1 = 1 \right\}$$

Note this is an n-dimensional analogue of the line-segment in two dimensions joining $(0, 1)$ and $(1, 0)$. It can be easily seen that $E$ is closed, bounded and convex. The condition defining $E$ can be expressed by means of the norm.

$$\|x\|_1 = \sum_{i=1}^{n} |x_i|$$

Then

$$E = \{x \in R^n: \|x\|_1 = 1 \text{ and } x_1 \geq 0 \text{ for each } i\}$$

It is a section of the unit sphere with respect to $\|.\|_1$. We are now in a position to project vectors onto $E$ : for any $x$ in the region $R^n$ where all components are non-negative, $\dfrac{x}{\|x\|_1} \in E$.

Define now an operator $T : E \rightarrow E$ by

$$Tx = \frac{Ax}{\|Ax\|_1} \text{ where } x \in E.$$

Since the elements of $A$ are positive, $\|Ax\|_1 > 0$ for all $x \in E$. Observe that the continuity of $A$ implies that of $T$. Applying Brouwer's fixed point theorem we conclude that

$$\frac{Ax}{\|Ax\|_1} = x$$

Hence, $x$ is an eigenvector of $A$, with $\|Ax\|_1$ as its eigenvalue, and the elements of $x$ are non-negative.

We have already seen that Brouwer's fixed point theorem is applicable to finite-dimensional spaces. In the following example we shall see that it does not apply to infinite-dimensional spaces.

**Example 6.2.2**  Consider the space $X$ of bilateral sequences such that

$$\sum_{-\infty}^{\infty} |x_n|^2 < \infty.$$

Denote

$$S = \{x \in X : \|x\| \leq 1\}$$

The set $S$ is closed, convex and bounded.

We shall produce a continuous operator $T : S \to S$ with no fixed point. Define an operator $U$ as follows

$$Ux = y$$

where

$$y_n = x_{n-1}$$

Clearly,

$$\|Ux\| = \|x\|$$

Define $T$ by

$$Tx = Ux + (1 - \|x\|)\, z$$

where $z$ is the sequence $\{z_n\}$ with $z_0 = 1$ and $z_n = 0$ for $n \neq 0$. It can be easily seen that $T$ is continuous. If $\|x\| \leq 1$, then

$$\|Tx\| \leq \| Ux \| + (1 - \|x\|) \|z\|$$

$$= \|x\| + 1 - \|x\| = 1$$

Hence, $T$ maps $S$ into $S$. We shall now show that $T$ has no fixed point. Suppose that $T$ has a fixed point $x$. Then

$$(Tx)_n = x_n$$

It now follows from the definition of $T$ that

$$x_n = \begin{cases} x_{n-1} & \text{if } n \neq 0 \\ x_{n-1} + 1 - \|x\| & \text{if } n = 0 \end{cases}$$

Hence,

$$x_{-1} = x_{-2} = \ldots \quad \text{and} \quad x_0 = x_1 = \ldots = 0.$$

Since $\sum_{-\infty}^{\infty} |x_n|^2 < \infty$, it follows that $x_n \to 0$ as $n \to \infty$. So the only possibility is $x_{-1} = x_{-2} = \ldots = 0$ and $x_0 = x_1 = \ldots = 0$. This implies that $\|x\| = 0$. This contradiction proves that $T$ has no fixed point.

## 6.3  SCHAUDER'S FIXED POINT THEOREM AND SOME RELATED RESULTS

G.D. Birkoff and O.D. Kellog first proved fixed-point theorems in infinite-dimensional spaces. They established the existence of fixed points for continuous self-maps on compact convex subsets of $C[0,1]$ and $L^2[0,1]$. *J.* Schauder generalized these results to normed linear space.

**Schauder's Theorem 6.3.1**  Let $E$ be a non-empty compact convex subset of a normed linear space $X$. Then every continuous map of $E$ into itself has a fixed point.

The proof is based on the deep topological idea as in Brouwer's theorem. In fact, the proof consists of approximating the infinite-dimensional set $E$ by a finite-dimensional set, applying Brouwer's theorem to deduce the existence of a fixed point of the finite-dimensional approximation, and then taking the limit as the dimension of the approximating space tends to infinity. We omit the proof.

We shall apply Schauder's theorem to integral equations. The following version of Schader's theorem is often useful.

**Theorem 6.3.2**  Let $E$ be a non-empty convex closed subset of a normed linear space $X$ and let $F$ be a relatively compact subset of $E$. Then every continuous mapping of $E$ into $F$ has a fixed point.

We now apply this theorem to integral equations of the form (non-linear)

$$u(x) = \int_a^b K(x,\, y)\, f(y,\, u(y))\, dy$$

Here $K$ and $f$ are given functions and $u$ is unknown. This equation is known as the Hammerstein equation.

**Existence Theorem for Hammerstein Equations**  Let $K(x,\, y)$ be continuous for $a \le x,\ y \le b$, and let $f(y,\, z), \dfrac{\partial}{\partial z} f(y,\, z)$ be continuous and bounded for $a \le y \le b$ and for all $z$. Then Hammerstein Equation has a continuous solution.

**Proof.**  Note that we shall consider the Banach space $C[a,\, b]$ and find a closed convex set in $C[a,\, b]$ which is mapped into a relatively compact subset of itself by the integral operator in the Hammerstein Equation. We then prove that the operator is continuous. We shall consider the sphere which is a convex set, and choose its radius so that it is mapped into itself.

Define an operator $T : C[a,\, b] \rightarrow C[a,\, b]$ by

$$(Tu)\,(x) = \int_a^b K(x,\, y)\, f(y,\, u(y))\, dy$$

Since $f$ is bounded, we have

$$|f(y,\, z)| \le M$$

and since $K$ is continuous on a compact set, it follows that

$$|K(x,\, y)| \le L$$

Hence,

$$(Tu)\,(x) \leq \int_a^b |k(x, y)| \, |f(y, u(y))| \, f(y, u(y)) \, dy$$

$$\leq LM \int_a^b dy = LM(b - a)$$

$$= \rho \ (\text{say})$$

Thus $T$ maps the whole space into the sphere of radius $\rho$. Write

$$E = \{u \in C[a, b] : \|u\| \leq \rho\}$$

Then $T : E \to E$ is not compact.

In order to apply Schader's Theorem we need to show that the set

$$F = \{Tu : u \in E\}$$

is relatively compact.

Observe that $F$ is uniformly bounded and $K$ uniformly continuous on the closed square $a \leq x, y \leq b$. It follows that for any $\epsilon > 0$ there is a $\delta > 0$ such that

$$|K(x_1, y) - K(x_2, y)| < \frac{\epsilon}{M(b - a)}$$

whenever $|x_1 - x_2| < \delta$. It also follows that

$$|(Tu)(x_1) - (Tu)(x_2)|$$

$$= \left| \int_a^b |K(x_1, y) - K(x_2, y)] \, f(y, u(y)) \, dy \right|$$

$$< (b - a) \, \frac{\epsilon}{M(b - a)} \, M = \epsilon$$

Thus $F$ is equicontinuous. By the Arzela-Ascoli theorem we find that $F$ is relatively compact.

Finally we need to show that $T$ is a continuous operator. Since $f$ has a bounded derivative,

$$|f(y, z_1) - f(y, z_2)| = \left| (z_1 - z_2) \, \frac{\partial}{\partial z} f(y, z) \right|$$

$$\leq C|z_1 - z_2|$$

for some $z$,

where

$$C = \sup \left| \frac{\partial}{\partial z} f(y, z) \right|. \quad \text{Hence, for any } \epsilon > 0,$$

$$|f(y, z_1) - f(y, z_2)| < \frac{\epsilon}{L(b - a)}$$

whenever $|z_1 - z_2| < \delta = \dfrac{\epsilon}{LC(b - a)}$

Thus

$$\|Tu_1 - Tu_2\| \le \int_a^b \left| K(x, y) \left[ f(y, u_1(y) - f(y, u_2(y))) \right] \right| dy$$

$$< (b - a) \frac{L\in}{L(b - a)} = \in$$

Provided $\|u_1 - u_2\| < \delta$. This proves that $T$ is continuous. By applying Schauder's fixed point Theorem we find that Hammerstein Equation has a continuous solution.

We now state the Banach fixed point theorem and some of its generalizations in tabular form.

| Banach | $X$ complete metric space | $f: X \to X$ contraction mapping | $f$ has a unique fixed point |
|---|---|---|---|
| | $X$ complete metric space | $f: X \to X$ continuous such that $f^k$ is contraction for some positive integer $k$ | $f$ has a unique fixed point |
| Edlestein 1962 | $X$ complete metric space | $f: X \to X$, $x_0 \in X$ such that $\{f^n x_0\}$ has a convergent subsequence converging to $u \in X$ | $u$ is unique fixed point of $f$ |
| Sehgal 1969 | $X$ metric space | $f: X \to X$ continuous $x$, $y \in X$, $x \ne y$, $d(f_x, f_y) <$ max $\{d(x, f_x), d(y, f_y), d(x, y)\}$. For some $z \in X$, $\{f^n z\}$ has a cluster point $u$. | $\{f^n z\}$ converges to $u$ and $u$ is a unique fixed point of $f$ |
| Byod and Wong 1969 | $X$ complete metric space | $f: X \to X$, $\varphi: R^+ \to R^+$ u.s.c. such that $d(f_x, f_y) \le \varphi(d(x, y))$, $\varphi(t) < t$ for $t > 0$. | $f$ has a unique fixed point $u \in X$ and for every $x \in X$, $f^n x \to u$. |
| Hardy and Regars 1973 | $X$ complete metric space | $f: X \to X$, $d(f_x, f_y) \le a [d(x, f_x) + d(y, f_y)] + b [d(y, f_x) + d(x, f_y)] + cd (x, y)$, $2a + 2b + c < 1$, $a, b, c \ge 0$ | $f$ has unique fixed point $f^n x \to u$. |
| Caristi 1976 | $X$ complete metric space | $g: X \to X$, $\varphi: X \to R^+$ l.s.c. such that $d(x, g_x) \le \varphi(x) - \varphi(g(x))$ for all $x \in X$ | $g$ has a fixed point |
| Downing and Kirk 1977 | $X, Y$ complete metric space | $g: X \to X$, $f: X \to Y$ closed, $\varphi: f(X) \to R^+$ l.s.c, $c > 0$ such that for each $x \in X$, $d(x, g_x) \le \varphi(f_x) - \varphi (g(f_x))$, $cd(f_x, f(g_x)) \le \varphi(f_x) - \varphi(g(f_x))$ | There exist $u \in X$ such that $g(u) = u$. |

*Contd...*

*Contd...*

| | X Banach space, *K* closed convex bounded subset of *X* | $f: K \to K$ nonexpansive, (id-$f$) ($K$) is a closed subset of *X* | $f$ has a fixed point |
|---|---|---|---|
| Browder 1965, Browder and Petryshyn 1967 | H Hilbert space, *K* closed convex bounded subset | $f: K \to K$ nonexpansive | $f$ has a fixed point |
| | X uniformly convex Banach space, *K* closed convex bounded subset | $f: K \to K$ nonexpansive | $f$ has a fixed point |
| Bose and Mukherjee 1981, Simi 1978 | X uniformly convex Banach space, *K* closed convex bounded | $f: K \to K$ continuous, $\|fx - fy\| \le a\|x - y\| + b[\|x - f_x\| + \|y - f_y\|] + c[\|x - f_y\| + \|y - f_x\|]$, $a, b, c \ge 0$, $a + 2b + 2c \le 1$ | $f$ has a fixed point |
| | X reflexive Banach spac, *K* closed bounded convex subset of *X*, *K* has a normal structure | $f: K \to K$ nonexpansive | $f$ has a fixed point |
| | X Banach space, *K* weakly compact convex subset of *X*, *K* has a normal structure | $f: K \to K$ nonexpansive | $f$ has a fixed point |
| | *H* a real Hilbert space, *K* closed bounded convex subset of *H* | $f: K \to H$ continuous, pseudo-contractive and weakly inward mapping. | $f$ has a fixed point |
| | X uniformly convex Banach space, *K* closed bounded convex subset of *X*. | $f: K \to K$ asymptotically nonexpansive | $f$ has a fixed point |

We now state below the Brower's and Schauder's fixed point theorems and some of their generalizations in a tabular form.

| | | | |
|---|---|---|---|
| Brower 1910 | *K* closed unit ball in $R^n$ | $f: K \to K$ continuous | $f$ has a fixed point |
| | *K* closed bounded (compact) convex subset of $R^n$ | $f: K \to K$ continuous | $f$ has a fixed point |
| Schauder 1927 | *K* compact convex subset of a normed space | $f: K \to K$ continuous | $f$ has a fixed point |

*Contd...*

*Contd...*

| | $K$ closed convex subset of a normed space and $F$ a relatively compact subset of $K$. | $f: K \rightarrow F$ continuous | $f$ has a fixed point |
|---|---|---|---|
| Tychonoff 1935 | $K$ compact convex subset of a locally convex Hausdorff topological vector space | $f: K \rightarrow K$ continuous | $f$ has a fixed point |
| Kakutani 1941 | $K$ closed bounded (compact) convex subset of $R^n$ | $f: K \rightarrow K$ upper semi-continuous multifunction with closed convex values | $f$ has a fixed point |
| Bobenblust and Karlin 1950 | $K$ compact convex subset of a Banach space | $f: K \rightarrow K$ use multifunction with closed convex values | $f$ has a fixed point |
| Kyfan and Glicksberg 1952 | $K$ compact convex subset of a locally convex space | $f: K \rightarrow K$ use multifunction with closed convex values | $f$ has a fixed point |
| Himmelberg 1966 | $K$ convex subset of a locally convex space | $f: K \rightarrow K$ upper semi-continuous compact with closed convex values | $f$ has a fixed point |

# Linear Operators

**7**

In this chapter we shall review the concept of bounded linear transformations.

**Definition 7.1.1**  Let $K$ denote the real field $R$ or the complex field $C$ and let $X$ and $Y$ be normed linear spaces over $K$. A linear transformation $T$ from $X$ into $Y$ is said to be a bounded linear transformation if there exists a constant $M > 0$ such that

$$\|Tx\| \leq M\|x\| \quad \text{for every } x \in X.$$

In the above inequality $\|x\|$ is the norm of $x$ in $X$ and $\|Tx\|$ is the norm of $Tx$ in $Y$. It will frequently happen that several norms occur together; but we will use the same symbol for all the norms. Seeing the nature of the discussion the reader can clearly distinguish between different norms.

If $T$ is a bounded linear transformation, the norm of $T$ is defined by

$$\|T\| = \sup \left\{ \frac{\|Tx\|}{\|x\|} : x \in X, \, x \neq 0 \right\}.$$

It may be noted that we can restrict ourselves only to $x \in X$ with $\|x\| = 1$ without changing the supremum, since for $\alpha \in K$,

$$\|T(\alpha x)\| = \|\alpha Tx\| = |\alpha| \, \|Tx\|$$

Therefore, the norm of $T$ can also be defined by

$$\|T\| = \sup \{\|Tx\| : x \in X, \|x\| = 1\}$$

Also, it may be noted that

$$\|T\| = \inf \{M : M > 0, \|Tx\| \leq M\|x\|\} \quad \text{for all } x \in X$$

In other words

$$\|Tx\| \leq \|T\|\|x\| \text{ for all } x \in X$$

If $T$ is a bounded linear transformation from a normed linear space $X$ into itself, then we call $T$ a bounded linear operator. Some authors also call any bounded linear transformation a bounded linear operator. A bounded linear transformation from $X$ into the field $K$ is called a bounded linear functional; it is called a real or complex bounded linear functional according as $K$ is the real field $R$ or the complex field $C$.

**Examples 7.1.1**

(a) The identity operator $I : X \to X$ on a normed linear space $X \neq \{0\}$ is a bounded linear operator with the norm $\|I\| = 1$.

(b) The zero transformation $0 : X \to Y$ on a normed linear space $X$ is a bounded linear transformation and has the norm $\|0\| = 0$

(c) The norm $\|\cdot\| : X \to R$ on a linear space $X$ is not a linear functional; it is a sublinear functional.

Other nontrivial examples will be discussed in due course.

The following theorem asserts that for a linear transformation, continuity and boundedness are equivalent.

**Theorem 7.1.1**    If $T$ is a linear transformation from a normed linear space $X$ into a normed linear space $Y$, then the following three conditions are equivalent :

(i) $T$ is bounded

(ii) $T$ is continuous

(iii) $T$ is continuous at one point.

**Proof.**    For a bounded linear transformation $T$ we have

$$\|Tx_1 - Tx_2\| = \|T(x_1 - x_2)\| \leq \|T\| \, \|x_1 - x_2\|.$$

Therefore (i) implies (ii). It is trivial that (ii) implies (iii). Now suppose $T$ is continuous at $x_0 \in X$. Then for each given $\in \, > 0$, there exists a $\delta > 0$ such that

$$\|x - x_0\| < \delta \Rightarrow \|Tx - Tx_0\| < \in$$

In other words, $\|x\| < \delta$ implies that

$$\|Tx\| = \|T(x_0 + x) - Tx_0\| < \in$$

Hence $\|T\| \leq \dfrac{\in}{\delta}$ and thus (iii) implies (i).

We have the following theorem.

**Theorem 7.1.2**    If $X$ and $Y$ are normed linear spaces, then $B(X, Y)$, the set of all bounded linear transformations from $X$ into $Y$, is a normed linear space with point wise linear operations and the operator norm. Moreover, $B(X, Y)$ is a Banach space if $Y$ is a Banach space.

**Proof.**    It is a routine verification that $B(X, Y)$ is a linear space. It is also trivial that $\|T\| = 0 \Leftrightarrow T = 0$ and $\|\lambda T\| = |\lambda| \, \|T\|$ for any $\lambda \in K$. Further, for $T_1$, $T_2 \in B(X, Y)$ we have

$$\|T_1 + T_2\| = \sup_{\|x\|=1} \|(T_1 + T_2)x\| = \sup_{\|x\|=1} \|T_1x + T_2x\|$$

$$\leq \sup_{\|x\|=1} \|T_1x\| + \sup_{\|x\|=1} \|T_2x\|$$

$$= \|T_1\| + \|T_2\|$$

Thus $B(X, Y)$ is a normed linear space. We need only to show that $B(X, Y)$ is complete if $Y$ is. To this end, let $\{T_n\}$ be a Cauchy sequence in $B(X, Y)$. For each $x \in X$, we have

$$\|T_n x - T_m x\| = \|(T_n - T_m)\, x\| \leq \|T_n - T_m\|\, \|x\|$$

so that $\{T_n x\}$ is a Cauchy sequence in $Y$ for each $x \in X$. Since $Y$ is complete, there is $Tx \in Y$ such that $Tx = \lim_{n \to \infty} T_n x$. Since each $T_n$ is linear, it follows that $T$ is linear. Further, since $\{T_n\}$ is a Cauchy sequence in a normed linear space, the sequence of numbers $\{\|T_n\|\}$ is bounded.

So

$$\|Tx\| = \|\lim T_n x\| = \lim \|T_n x\|$$

$$< \sup(\|T_n\|\, \|x\|) = (\sup \|T_n\|)\, \|x\|$$

shows that $T$ is bounded. Thus $T \in B\,(X, Y)$. It remains to show that

$$\lim_{n \to \infty} \|T_n - T\| = 0.$$

Let $\epsilon > 0$ be given. Then there exists a positive integer $n_0$ such that

$$\|T_n - T_m\| < \epsilon \text{ for } n, m \geq n_0.$$

If $\quad \|x\| = 1$, and $m, n \geq n_0$, then

$$\|T_n x - T_m x\| = \|(T_n - T_m)\, x\| \leq \|T_n - T_m\|\, \|x\|$$

$$= \|T_n - T_m\| < \epsilon$$

Let $n$ be fixed and let $m \to \infty$. Then we have

$$\|T_n x - T_m x\| \to \|T_n x - Tx\|,$$

from which it follows that $\|T_n x - Tx\| \leq \epsilon$ for all $n \geq n_0$ and all $x$ such that $\|x\| = 1$. This shows that $\|T_n - T\| \leq \epsilon$ for all $n \geq n_0$ and this completes the proof.

*Remark 7.1.1*    It follows from the above theorem that $B(X)$, the set of all bounded linear operators defined on a normed linear space $X$ is a normed linear space and that $B(X)$ is a Banach space if $X$ is a Banach space.

Also $X^*$, the set of all bounded linear functionals defined on a normed linear space $X$ is a Banach space. It should be noted that $X^*$ is always a Banach space even if $X$ is not complete.

## 7.1 BOUNDED LINEAR OPERATORS ON HILBERT SPACES

In this section we shall discuss some elementary results on bounded linear operators in Hilbert spaces. We shall first prove a theorem which is a direct application of the Riesz representation theorem. Throughout this section $B(X)$ denotes the set of all bounded linear operators on a Hilbert space $X$.

**Theorem 7.1.3**   If $T \in B(X)$, then there exists a unique $U \in B(X)$ such that

$$(Tx, y) = (x, Uy) \quad \text{for } x, y \in X.$$

**Proof.**   For a fixed $y \in X$, let $\phi : X \to C$ be defined by

$$\phi(x) = (Tx, y) \text{ for all } x \in X.$$

It is easy to see that $\phi$ is a bounded linear functional on $X$. Therefore by Riesz representation theorem, there exists a unique $z \in X$, such that $\phi(x) = (x, z)$ for all $x \in X$. Define $Uy = z$. Clearly $U$ is linear and $(Tx, y) = (x, Uy)$. Putting $x = Uy$ we have

$$\|Uy\|^2 = (Uy, Uy) = (TUy, y) \leq \|T\| \, \|Uy\| \, \|y\|$$

for all $y \in X$. Therefore $\|U\| \leq \|T\|$, i.e., $U$ is bounded and hence $U \in B(X)$.

Suppose there is another $V \in B(X)$ satisfying $(x, Vy) = (Tx, y)$. Then $(x, Uy - Vy) = 0$. This implies $Uy - Vy = 0$. Hence $U = V$ and this completes the proof.

This theorem guarantees the existence of the adjoint.

**Definition 7.1.2**   Let $T \in B(X)$. Then the adjoint of $T$, denoted by $T^*$, is the unique element of $B(X)$ satisfying $(Tx, y) = (x, T^*y)$ for all $x, y \in X$.

**Theorem 7.1.4**   The mapping $T \to T^*$ of $B(X)$ into itself has the following properties: For $T, T_1, T_2 \in B(X)$ and $\alpha \in C$ we have

(i) $(T_1 + T_2)^* = T_1^* + T_2^*$;

(ii) $(\alpha T)^* = \overline{\alpha} T^*$;

(iii) $(T_1 T_2)^* = T_2^* T_1^*$;

(iv) $T^{**} = (T^*)^* = T$;

(v) $\|T^*\| = \|T\|$;

(vi) $\|T^* T\| = \|T\|^2$.

**Proof.**

(i) For $x, y \in X$,

$$((T_1 + T_2)\, x,\, y) = (T_1 x,\, y) + (T_2 x,\, y)$$
$$= \left(x,\, T_1^* y\right) + \left(x,\, T_2^* y\right)$$
$$= (x, (T_1^* + T_2^*) y)$$

(ii) For $x, y \in X$,

$$((\alpha T)\, x,\, y) = (\alpha T x,\, y) = \alpha (T x,\, y)$$
$$= \alpha (x,\, T^* y) = (x,\, \overline{\alpha} T^* y)$$

(iii) For $x, y \in X$,

$$(T_1\, T_2\, x,\, y) = (T_1 (T_2 x),\, y) = \left(T_2 x,\, T_1^* y\right)$$
$$= \left(x,\, T_2^*\, T_1^* y\right)$$

(iv) For $x, y \in X$,

$$(x,\, T^{**} y) = (T^* x,\, y) = \overline{\left(y,\, T^* x\right)}$$
$$= \overline{(\overline{T y,\, x})} = (x,\, T y)$$

(v) We have $\|T^*\| \leq \|T\|$. So

$$\|T\| = \|T^{**}\| \leq \|T^*\| \leq \|T\|$$

which shows that $\|T\| = \|T^*\|$.

(vi) First note that for $T_1, T_2 \in B(X)$ and $x \in X$ we have

$$\|(T_1\, T_2)\, x\| = \|T_1\, (T_2 x)\| \leq \|T_1\| \, \|T_2 x\| \leq \|T_1\| \, \|T_2\| \, \|x\|$$

and hence $\|T_1\, T_2\| \leq \|T_1\| \, \|T_2\|$. Therefore it follows that

$$\|T^* T\| \leq \|T^*\| \, \|T\| = \|T\|^2$$

For $x \in X$ we have

$$\|T x\|^2 = (T x,\, T x) = (x,\, T^* T x)$$
$$\leq \|x\| \, \|T^* T x\| \leq \|x\|^2 \, \|T^* T\|$$

Therefore $\qquad \|T x\| \leq \|T^* T\|^{1/2} \, \|x\|$

Thus $\qquad \|T\| \leq \|T^* T\|^{1/2}$ and hence $\|T\|^2 \leq \|T^* T\|$

Therefore $\|T^* T\| = \|T\|^2$ and this completes the proof.

We now introduce some important classes of operators.

**Definition 7.1.3**   Let $T \in B(X)$. Then

   (i) $T$ is self-adjoint or hermitian if $T = T^*$,

   (ii) $T$ is normal if $T^*T = TT^*$,

   (iii) $T$ is unitary if $T^*T = TT^* = 1$ (identity operator),

   (iv) $T$ is positive if $(Tx, x) \geq 0$     for all $x \in X$.

**Remark 7.1.2**   For any $T \in B(X)$. $T^*T$ is a positive operator.

   For $x \in X$ we have

$$(T^*Tx, x) = (Tx, Tx) = \|Tx\|^2 \geq 0$$

and hence the result follows.

**Theorem 7.1.5**   $T \in B(X)$ is self-adjoint if and only if $(Tx, x)$ is a real number for every $x \in X$. In particular, every positive operator is self-adjoint.

**Proof.**   Let $T$ be self-adjoint and let $x \in X$. Then

$$(Tx, x) = (x, T^*x) = (x, Tx) = (\overline{Tx, x}).$$

Hence $(Tx, x)$ is a real number. Conversely, let $(Tx, x)$ be a real number. We have

$$4(Tx, y) = (T(x + y), x + y) - (T(x - y), x - y)$$
$$+ i(T(x + iy), x + .iy) - .i(T(x - .iy), (x - iy))$$

Taking conjugates and interchanging $x$ and $y$ we get

$$4(x, Ty) = (T(x + y), x + y) - (T(x - y), x - y)$$
$$-i(T(y + ix), y + ix) + i(T(y - ix), (y - ix))$$

Comparing the right side of the above two equations we see that

$$(Tx, y) = (x, Ty)$$

which shows that $T = T^*$ and therefore $T$ is self-adjoint.

**Remark 7.1.3**   If $T_1$ and $T_2$ are self-adjoint, then $T_1T_2$ is self-adjoint if and only if

$$T_1T_2 = T_2T_1.$$

**Proof.**   Observe that

$$(T_1 T_2)^* = T_2^* T_1^* = T_2T_1$$

**Theorem 7.1.6**   The set of all self-adjoint operators is a closed (real) linear subspace of $B(X)$ and therefore is a real Banach space.

**Proof.** Clearly $O$ and $I$ are self-adjoint. If $T_1$ and $T_2$ are self-adjoint and $\alpha$ and $\beta$ are real numbers, then

$$(\alpha T_1 + \beta T_2)^* = \overline{\alpha} T_1^* + \overline{\beta} T_2^* = \alpha T_1 + \beta T_2$$

so that $\alpha T_1 + \beta T_2$ is self-adjoint. If $\{T_n\}$ is a sequence of self-adjoint operators which converges to $T$, then

$$\|T - T^*\| \leq \|T - T_n\| + \|T_n - T_n^*\| + \|T_n^* - T^*\|$$

$$= \|T - T_n\| + \|(T_n - T)^*\|$$

$$= \|T - T_n\| + \|T_n - T\|$$

$$= 2\|T_n - T\| \to 0 \text{ as } n \to \infty$$

Therefore $T = T^*$. Thus $T$ is self-adjoint and this completes the proof.

**Theorem 7.1.7** The set of all normal operators is a closed subset of $B(X)$ which contains the set of all self-adjoint operators and is closed under scalar multiplication.

**Proof.** Obviously every self-adjoint operator is normal and if $T$ is normal and $\alpha$ is any scalar, then $\alpha T$ is normal. If $\{T_n\}$ is a sequence of normal operators such that $T_n \to T$, then $T_n^* \to T^*$ and so

$$\|TT^* - T^*T\| \leq \|TT^* - T_n T_n^*\| + \|T_n T_n^* - T_n^* T_n\|$$

$$+ \|T_n^* T_n - T^*T\|$$

$$= \|TT^* - T_n T_n^*\| + \|T_n^* T_n - T^*T\|$$

$$\to 0 \text{ as an } n \to \infty$$

This implies that $TT^* = T^*T$ and hence $T$ is normal. This completes the proof.

**Theorem 7.1.8** If $T_1$ and $T_2$ are normal operators such that each commutes with the adjoint of the other, then $T_1 + T_2$ and $T_1 T_2$ are normal.

**Proof.** We have

$$T_1 T_2^* = T_2 T_1^* \Leftrightarrow T_1 T_2^* = T_2^* T_1$$

So the assumption really implies that each commutes with the adjoint of the other. Also

$$(T_1 + T_2)(T_1 + T_2)^* = (T_1 + T_2)\left(T_1^* + T_2^*\right)$$

$$T_1 T_1^* + T_1 T_2^* + T_2^* T_1 + T_2 T_2^*$$

and

$$(T_1 + T_2)^* (T_1 + T_2) = (T_1^* + T_2^*)(T_1 + T_2)$$
$$= T_1^* T_1 + T_1^* T_2 + T_2^* T_1 + T_2^* T_2$$

The result that $T_1 + T_2$ is normal follows from above two equations. Similarly, the fact that $T_1 T_2$ is normal follows from

$$T_1 T_2 (T_1 T_2)^* = T_1 T_2 T_2^* T_1^* = T_1 T_2^* T_2 T_1^*$$
$$= T_2^* T_1 T_1^* T_2 = T_2^* T_1^* T_1 T_2 = (T_1 T_2)^* T_1 T_2$$

**Theorem 7.1.9** $T \in B(X)$ is normal if and only if its real and imaginary parts commute.

**Proof.** If $A_1$ and $A_2$ are the real and imaginary parts of $T$, then

$$T = A_1 + .iA_2 \text{ and } T^* = A_1 - .iA_2.$$

$$TT^* = (A_1 + iA_2)(A_1 - iA_2) = A_1^2 + A_1^2 + i(A_2 A_1 - A_1 A_2)$$

and $$T^* T = (A_1 - iA_2)(A_1 + iA_2) = A_1^2 + A_2^2 + i(A_1 A_2 - A_2 A_1)$$

Now it is evident that if $A_1 A_2 = A_2 A_1$ then $TT^* = T^*T$. Conversely, if $TT^* = T^*T$, then $A_1 A_2 - A_2 A_1 = A_2 A_1 - A_1 A_2$, so $A_1 A_2 = A_2 A_1$.

To obtain some further results on normal operators we first prove

**Theorem 7.1.10** If $T \in B(X)$ is such that $(Tx, x) = 0$ for all $x$, then

$$T = 0.$$

**Proof.** We need only to show that $(Tx, y) = 0$ for all $x, y \in X$. An easy computation shows that

$$(T(\alpha x + \beta y), \alpha x + \beta y) - |\alpha|^2(Tx, x) - |\beta|^2(Ty, y)$$
$$= \alpha\bar{\beta}(Tx, y) + \bar{\alpha}\beta(Ty, x)$$
$$= 0 \text{ (by the hypothesis).}$$

If $\alpha = 1$ and $\beta = 1$ we get

$$(Tx, y) + (Ty, x) = 0$$

and if $\alpha = i$ and $\beta = 1$ we have

$$(Tx, y) - (Ty, x) = 0$$

Therefore, $(Tx, y) = 0$ and this completes the proof.

**Theorem 7.1.11** (i) $T$ is normal $\Leftrightarrow \|T^* x\| = \|Tx\|.$

(ii) If $T$ is normal, then $\|T^2\| = \|T\|^2.$

**Proof.**  (i) We have

$$\|T^*x\| = \|Tx\| \Leftrightarrow \|T^*x\|^2 = \|Tx\|^2$$

$$\Leftrightarrow (T^*x, T^*x,) = (Tx, Tx)$$

$$\Leftrightarrow (TT^*x, x) = (T^*Tx, x)$$

$$\Leftrightarrow ((TT^* - T^*T)\, x, x) = 0$$

Hence the result follows.

(ii) If $T$ is normal, then from (i) we have

$$\|T^2x\| = \|TTx\| = \|T^*Tx\|$$

for every $x$ and hence

$$\|T^2\| = \|T^*T\| = \|T\|^2$$

**Theorem 7.1.12**  $T \in B(X)$ is unitary if and only if it is an isometric isomorphism of $X$ onto itself.

**Proof.**  Let $T$ be unitary. Then $(Tx, Ty) = (T^*Tx, y) = (x, y)$ for all $x, y \in X$.

Therefore $\|Tx\|^2 = (Tx, Tx) = (x, x) = \|x\|^2$, i.e. $\|Tx\| = \|x\|$. Thus $T$ preserves the norm, it is clearly onto and hence is an isometric isomorphism.

Let $T$ be an isometric isomorphism of $X$ onto itself. Then

$$(x, x) = \|x\|^2 = \|Tx\|^2 = (Tx, Tx) = (T^*Tx, x)$$

So $((T^*T - 1)\, x, x) = 0$ and thus it follows that $T^*T = 1$. Also $T^{-1}$ exists and

$$T^* = (T^*T)\, T^{-1} = 1T^{-1}, \text{ so } TT^* = TT^{-1} = 1. \text{ Thus}$$

$$T^*T = TT^* = 1.$$

Hence $T$ is unitary and this completes the proof.

**Theorem 7.1.13**  Let $B(X)$ denote the set of all bounded linear operators on a Banach space $X$. Let $T \in B(X)$ and $\|I - T\| < 1$

$$\|T^{-1}\| \le \frac{1}{1 - \|I - T\|}$$

**Proof.**  For $N \ge M$ we have

$$\left\| \sum_{n=0}^{N} (1 - T)^n - \sum_{n=0}^{M} (1 - T)^n \right\| = \left\| \sum_{n=M+1}^{N} (1 - T)^n \right\|$$

$$\le \sum_{n=M+1}^{N} \|1 - T\|^n \to 0$$

as $N$, $M \to \infty$ since $\|1 - T\| < 1$. Therefore the sequence of partial sums $\left\{ \sum_{n=0}^{N} (1 - T)^n \right\}_{N=0}^{\infty}$ is a Cauchy sequence. Since $B(X)$ is complete we have

$$\sum_{n=0}^{\infty} (1 - T)^n = U \text{ (say)}$$

Then

$$TU = [I - (I - T)] \left[ \sum_{n=0}^{\infty} (1 - T)^n \right]$$

$$= \lim_{N \to \infty} \left( [I - (I - T)] \sum_{n=0}^{N} (I - T)^n \right)$$

$$= \lim_{N \to \infty} \left( I - (I - T)^{N+1} \right) = 1$$

since $\lim_{N \to \infty} \left\| (1 - T)^{N+1} \right\| = 0$. Similarly it can be shown that $UT = I$ and so $T^{-1} = U$. Further

$$\|v\| = \lim_{N \to \infty} \left\| \sum_{n=0}^{N} (1 - T)^n \right\| \le \lim_{N \to \infty} \sum_{n=0}^{N} \left\| 1 - T \right\|^n$$

$$= \frac{1}{1 - \|I - T\|}$$

Let $G$ denote the set of all invertible operators (i.e., the operators which have multiplicative inverse in $B(X)$.

**Theorem 7.1.14**   $G$ is an open set. Further, the map $T \to T^{-1}$ of $G$ into $G$ is continuous and is therefore a homeomorphism of $G$ onto itself.

**Proof.**   Let $T \in G$ and let $U \in B(X)$ be such that $\|U - T\| < \dfrac{1}{\|T^{-1}\|}$. Now

$$\|T^{-1} U - I\| = \|T^{-1}(U - T)\| \le \|T^{-1}\| \, \|U - T\| < 1$$

and hence

$$\|U^{-1}\| = \|(U^{-1}T)T^{-1}\| \le \|U^{-1}T\|\|T^{-1}\|$$

$$= \|(T^{-1} U)^{-1}\| \, \|T^{-1}\| < 2 \, \|T^{-1}\|$$

Thus $\quad \|T^{-1} - U^{-1}\| = \|T^{-1}(T - U)^{-1}\| \le \| T^{-1}\|\|T - U\|\|U^{-1}\|$

$$\le 2\| T^{-1}\|2\|T - U\|$$

This shows that $T \to T^{-1}$ is continuous and completes the proof.

## 7.2 EIGEN VALUES

**Definition 7.2.1** Let $X$ be a vector space and let $T : X \to X$ be linear. The number $\lambda$ is called an eigen value of $T$ if there is a non-zero $u \in X$ such that

$$Tu = \lambda u$$

Any such non-zero $u$ is called an eigen vector of $T$, and is said to correspond to the eigen value $\lambda$. If the elements of $X$ are functions, $u$ is called an eigen function.

We shall now give an illustration from mechanics.

In the theory of vibrating system, consider the following situation.

The state of a system at any given time $t$ may be represented by an element $u(t)$ of a Hilbert space $H$, and the equation of motion is of the form

$$\frac{d^2 u}{dt^2} = Tu$$

where $T$ is an operator on $H$. When the system vibrates, the time-dependence of $u$ will be sinusoidal;

$$u(t) = \sin \omega t)v$$

where $v$ is a fixed element of Hilbert space $H$. Suppose that $T$ is linear, then the equation of motion becomes

$$-\omega^2 v = Tv$$

This means that $-\omega^2$ is an eigen value of $T$. In other words, the eigen values of the operator $T$ correspond to the possible frequencies of vibration.

**Definition 7.2.2** The set of all eigen vectors belonging to one particular eigen value of an operator is called the eigen space of that eigen value.

**Example 7.2.1** Define $T : L_2 [a, b] \to L_2 [a, b]$ by

$$(Tf)\,(x) = \int_a^b K(x, y)f(y)dy \qquad \qquad \ldots(7.1)$$

where $K$ is a continuous function. Take $K(x, y) = \cos(x - y)$ and $a = 0$, $b = 2\pi$. Then the eigen value equation Eq. (7.1) becomes

$$\int_0^{2\pi} \cos(x - y)\, u(y)dy = \lambda u(x)$$

i.e.

$$\cos x \int_0^{2\pi} \cos y\, u(y)dy + \sin x \int_0^{2\pi} \sin y\, u(y)dy$$

$$= \lambda u(x) \qquad \qquad \ldots(7.2)$$

Suppose $\lambda \neq 0$. Then the equation (7.2) implies that $u(x)$ is a linear combination of $\sin x$ and $\cos x$ with constant coefficients. Thus for $\lambda \neq 0$,

$$u(x) = a \cos x + b \sin x \qquad \ldots(7.3)$$

for some constants a and b. Substituting this in (7.2) we find that

$$\pi a = \lambda a, \ \pi b = \lambda b \qquad \ldots(7.4)$$

Thus $\lambda = \pi$, and $a$ and $b$ can take any values. It follows that this operator has exactly one non-zero eigen value $\pi$, and its eigen functions are of the form $a \cos x + b \sin x$.

Observe that (7.2) shows that zero is also an eigen value, its eigen functions being all functions orthogonal to $\cos y$ and $\sin y$; it has infinite multiplicity.

**Theorem 7.2.1**    Let $T$ be a self-adjoint operator on a Hilbert space $H$. Then all its eigen values are real, and eigen vectors corresponding to different eigen values are orthogonal.

**Proof.**    Suppose that $Tu = \lambda u$. Then

$$\lambda \|u\|^2 = \lambda \, (u, u) = (Tu, u)$$

$$= (u, Tu) = (u, \lambda u) = \overline{\lambda} \|u\|^2$$

Since $u \neq 0$, it follows that $\lambda = \overline{\lambda}$, so $\lambda$ is real.

Suppose now that $Tv = \mu v$. Then

$$(v, Tu) = (Tv, u)$$

$$\lambda(v, u) = \mu(v, u)$$

**Theorem 7.2.2**    Let $T$ be a bounded operator in a Hilbert space $H$. If $\lambda$ is an eigen value of $T$, then

$$|\lambda| \leq \|T\|$$

**Proof.**    Suppose $Tu = \lambda u$. Then

$$\|\lambda u\| = \|Tu\|$$

Hence $\qquad |\lambda|\|u\| \leq \|T\|\|u\|$

It follows that $|\lambda| \leq \|T\|$

Since the eigen values are points in the complex plane, this result means that all eigen values lie inside a circle of radius $\|T\|$.


## 7.3  SPECTRUM

In this section we shall discuss the concept of spectrum of an element of $B(X)$. The spectrum is a subset of the complex field which reflects the algebraic properties of $B(X)$ and does not depend on the norm.

We shall now introduce the definition of the spectrum and prove that it is compact.

**Definition 7.3.1** Let $T \in B(X)$. The spectrum of $T$ is the set $\sigma(T)$ defined by

$$\sigma(T) = \{\lambda \in C : T - \lambda \text{ is not invertible}\}$$

The resolvent set of $T$, denoted by $\rho(T)$, is the complement of the spectrum in the complex plane. The spectral radius of $x$ is defined by

$$r(T) = \sup \{|\lambda| : \lambda \in \sigma(T)\}.$$

The concept of spectrum and resolvent set can also be defined as follows:

Let $X \neq \{0\}$ be complex normed space and $T : D(T) \to X$ a linear operator with domain $D(T) \subset X$. With $T$ we associate the operator.

(1)  $T_\lambda = T - \lambda I$

where $\lambda$ a complex number and $I$ is the identity operator on $D(T)$. If $T_\lambda$ has an inverse, we denote it by $R_\lambda(T)$, that is,

(2)  $R_\lambda(T) = T_\lambda^{-1} = (T - \lambda I)^{-1}$

**Definition 7.3.2** Let $X \neq \{0\}$ be a complex normed space and $T : D(T) \to X$ a linear operator with domain $D(T) \subset X$. A regular value $\lambda$ of $T$ is a complex number such that

(R1)           $R_\lambda(T)$   exists,

(R2)           $R_\lambda(T)$   is bounded,

(R3)           $R_\lambda(T)$   is defined on a set which is dense in $X$.

The resolvent set $\rho(T)$ of $T$ is the set of all regular values $\lambda$ of $T$. Its complement $\sigma(T) = C - \rho(T)$ in the complex plane $C$ is called the spectrum of $T$, and a $\lambda \in \sigma(T)$ is called a spectral value of $T$. Furthermore, the spectrum $\sigma(T)$ is partitioned into three disjoint sets as follows.

The **point spectrum** or discrete spectrum $\sigma_p(T)$ is the set such that $R_\lambda(T)$ does not exist. A $\lambda \in \sigma_p(T)$ is called an eigen value of $T$.

The **continuous spectrum** $\sigma_c(T)$ is the set such that $R_\lambda(T)$ exists and satisfies (R3) but not (R2), that is, $R_\lambda(T)$ is unbounded.

The **residual spectrum** $\sigma_r(T)$ is the set such that $R_\lambda(T)$ exists (and may be bounded or not) but does not satisfy (R3), that is, the domain of $R_\lambda(T)$ is not dense in $X$.

Some of the sets in this definition may be empty. This is an existence problem which we shall have to discuss for instance, $\sigma_c(T) = \sigma_r(T) = \theta$ in the finite dimensional case.

The conditions stated can be summarized in the following table.

| Satisfied | Not satisfied | $\lambda$ belongs to |
|---|---|---|
| (R1), (R2), (R3) | | $\rho(T)$ |
| (R1)      (R3) | (R1) | $\sigma_p(T)$ |
| (R1) | (R2) | $\sigma_c(T)$ |
| | (R3) | $\sigma_r(T)$ |

We first note that the four sets in the table are disjoint and their union is the whole complex plane:

Hence if $R_\lambda x = (T - \lambda I)\, x = 0$ for some $x \neq 0$, then $\lambda \in \sigma_p(T)$, by definition, that is, $\lambda$ is an eigen value of $T$. The vector $x$ is then called an eigen vector of $T$ (or eigen function of $T$ if $X$ is a function space) corresponding to the eigen value $\lambda$. The subspace of $D\,(T)$ consisting of 0 and all eigen vectors of $T$ corresponding to an eigen value $\lambda$ of $T$ is called the eigen space of $T$ corresponding to that eigen value $\lambda$.

We see that our definition of an eigen value is in harmony with that in the preceding section. We also see that the spectrum of a linear operator on a finite dimensional space is a pure point spectrum, that is both the continuous spectrum and the residual spectrum are empty, as was mentioned before, so that every spectral value is an eigen value.

If $X$ is infinite dimensional, then $T$ can have spectral values which are not eigen values.

**Example (operator with a spectral value which is not an eigen value)**

On the Hilbert sequence space $X = l^2$ we define a linear operator $T : l_2 \to l_2$ by

$$(\xi_1, \xi_2, \ldots.) \to (0, \xi_1, \xi_2, \ldots.),$$

where $x = (\xi_j) \in l^2$. The operator $T$ is called the right-shift operator. $T$ is bounded (and $\|T\| = 1$) because

$$\|Tx\|^2 = \sum_{j=1}^{\infty} |\xi_j|^2 = \|x\|^2.$$

The operator $R_0(T) = T^{-1}$ exists; in fact, it is the left-shift operator given by

$$(\xi_1, \xi_2, \ldots.) \to (0, \xi_1, \xi_2, \ldots.),$$

But $R_0(T)$ does not satisfy (R3), because (3) shows that $T(X)$ is not dense in $X$; indeed, $T(X)$ is the subspace $Y$ consisting of all $y = (\eta_j)$ with $\eta_1 = 0$.

Hence, by definition, $\lambda = 0$ is a spectral value of $T$. Furthermore, $\lambda = 0$ is not an eigen value. We can see this directly since $Tx = 0$ implies $x = 0$ and the zero vector is not an eigen vector.

**Theorem 7.3.1**   Let $T \in B(X)$. Then $\sigma(T)$ is compact and $r(T) \leq \|T\|$.

**Proof.**   Let $\phi : C \to B(X)$ be defined by

$$\phi(\lambda) = T - \lambda$$

Then clearly $\phi$ is continuous and $\rho(T) = \phi^{-1}(G)$ where $G$ denotes the set of all invertible elements. Since $G$ is open, $\rho(T)$ is open and therefore the set $\sigma(T)$ is closed. Further if $|\lambda| > \|T\|$, then

$$\left\| 1 - \left(1 - \frac{T}{\lambda}\right) \right\| = \left\| \frac{T}{\lambda} \right\| = \frac{\|T\|}{|\lambda|} < 1$$

so that $1 - \dfrac{T}{\lambda}$ is invertible. Thus $T - \lambda$ is invertible and hence $\lambda \notin \sigma(T)$. Therefore $\sigma(T)$ is bounded and $r(T) \leq \|T\|$. $\sigma(T)$, being closed and bounded, is compact and this completes the proof.

**Theorem 7.3.2**   Let $T \in B(X)$, then $\sigma(T) \neq \phi$.

**Proof.**   Let $F : \rho(T) \to B(X)$ be defined by

$$F(\lambda) = (T - \lambda)^{-1}$$

Since inversion is continuous we have for $\lambda_0 \in \rho(T)$,

$$\lim_{\lambda \to \lambda_0} \frac{F(\lambda) - F(\lambda_0)}{\lambda - \lambda_0} = \lim_{\lambda \to \lambda_0} \frac{(T - \lambda_0)^{-1}[(T - \lambda_0) - (T - \lambda)](T - \lambda)^{-1}}{\lambda - \lambda_0}$$

$$= \lim_{\lambda \to \lambda_0} (T - \lambda_0)^{-1} (T - \lambda)^{-1} = (T - \lambda_0)^{-2}$$

Thus for $\phi \in (X^*)$ the function $\phi(F)$ is a complex analytic function on $\rho(T)$.

Further, for $|\lambda| > \|T\|$, $1 - \dfrac{T}{\lambda}$ is invertible and

$$\left\| \left(1 - \frac{T}{\lambda}\right)^{-1} \right\| \leq \frac{1}{1 - \left\| \dfrac{T}{\lambda} \right\|}$$

Thus, it follows that

$$\lim_{\lambda \to \infty} \|F(\lambda)\| = \lim_{\lambda \to \infty} \left\| \frac{1}{\lambda}\left(\frac{T}{\lambda} - 1\right)^{-1} \right\|$$

$$\leq \lim_{|\lambda| \to \infty} \sup \frac{1}{|\lambda|} \cdot \frac{1}{1 - \left\|\dfrac{T}{\lambda}\right\|} = 0$$

Therefore, for $\phi \in X^*$, we have $\lim_{\lambda \to \infty} \phi(F(\lambda)) = 0$.

Assume that $\sigma(T) = \phi$, so that $\rho(T) = C$. Thus for $\phi \in X^*$ it follows that $\phi(F)$ is an entire function that vanishes at infinity. By Liouville's theorem we have $\phi\ (F) \equiv 0$. Since for a fixed $\lambda \in C$, $\phi(F(\lambda)) = 0$ for each $\phi \in X^*$, it follows that $F(\lambda) = 0$. This is a contradiction since $F(\lambda)$, by definition, is an invertible element of $B(X)$. Therefore $\sigma(T) \neq \phi$ and this completes the proof.

The following theorem gives the formula for the spectral radius.

**Theorem 7.3.3**   (Spectral radius formula) Let $T \in B\ (X)$. Then

$$r(T) = \lim_{n \to \infty} \|T^n\|^{1/n}$$

**Proof.**   Let $T \in B\ (X)$, n any positive integer and $\lambda$ a complex number. Assume that $\lambda^n \notin \sigma(T^n)$. We have

$$(T^n - \lambda^n) = (T - \lambda)\left(T^{n-1} + \lambda T^{n-2} + \ldots + \lambda^{n-1}\right)$$

Multiplying both sides by $(T^n - \lambda^n)$, it follows that $T - \lambda$ is invertible and hence $\lambda \notin \sigma\ (T)$.

So if $\lambda \notin \sigma\ (T)$, then $\lambda^n \in \sigma(T^n)$ for $n = 1, 2, \ldots$ Therefore

$$|\lambda^n| \leq \|T^n\|$$

and hence $|\lambda| \leq \|T^n\|^{1/n}$.

This gives

$$r(T) \leq \liminf_{n} \|T^n\|^{1/n}$$

Now, consider the analytic function

$$G(\lambda) = -\lambda \sum_{n=0}^{\infty} \frac{T^n}{\lambda^n}$$

which converges for $|\lambda| > \lim_{n} \sup \|T^n\|^{1/n}$. For $|\lambda| > \|T\|$ we have $G\ (\lambda) = (T - \lambda)^{-1}$ and, therefore,

$$(T - \lambda)G(\lambda) = G(\lambda)(T - \lambda) = I \text{ for } |\lambda| > \lim_{n} \sup \|T^n\|^{1/n}$$

Thus $r(T) \geq \lim_{n} \sup \|T^n\|^{1/n} \geq \lim_{n} \inf \|T^n\|^{1/n} \geq r(T)$ and this completes the proof.

**Theorem 7.3.4**     $r(T) = \|T\|$ *iff* $\|T^2\| = \|T\|^2 \cdot$

In particular, if $T$ is normal,

$$r(T) = \|T\| \cdot$$

**Proof.**  If $r(T) = \|T\|$ clearly

$$\|T^2\| = r(T^2) = (r(T))^2 = \|T\|^2 \cdot$$

Conversely, if $\|T^2\| = \|T\|^2$,
for every positive integer $k$,

$$\left\|T^{2^k}\right\| = \|T\|^{2^k}.$$

Now

$$r(T) = \lim\|T^n\|^{1/n} = \lim\left\|T^{2^k}\right\|^{1/2k}$$

$$= \lim\|T\| = \|T\| \cdot$$

Spectrum of the shift operator on $l_2(Z)$:

$$l_2(Z) = \left\{ x : \sum_{n=-\infty}^{\infty} |x_n|^2 < \infty \right\}$$

$$U : l_2(Z) \to l_2(Z)$$

$$(Ux)_n = x_{n-1}.$$

$$\|Ux\|_2^2 = \sum_{n=-\infty}^{\infty} |(Ux)_n|^2 = \sum_{n=-\infty}^{\infty} |x_{n-1}|^2$$

$U$ preserves norm $\Rightarrow$ U is bounded.

$$A : l_2(Z) \to l_2(Z)$$
$$(Ax)_n = x_{n+1}$$

$$(Ux, y) = \sum_{n=-\infty}^{\infty} (Ux)_n \bar{y}_n = \sum_{n=-\infty}^{\infty} x_{n-1} \bar{y}_n$$

$$\sum_{n=-\infty}^{\infty} x_n \bar{y}_{n+1} = (x, Ay)$$

$$\Rightarrow \qquad A = U^*$$
$$UU^* = U^*U = I \quad \text{or} \quad U^{-1} = U^* = A$$

$$\Rightarrow \quad U \text{ is unitary.}$$

$$\sigma(U) \subset \overline{D} \text{ and } \sigma(U^{-1}) = \sigma(U^*) \subset \overline{D}.$$

For $\qquad \lambda \in \overline{D},\ 0 < |\lambda| < 1$

$$(U - \lambda)U^{-1} = UU^{-1} - \lambda U^{-1} = I - \lambda\ U^{-1}$$

$$= \left(\frac{1}{\lambda} - U^{-1}\right)$$

$\dfrac{1}{\lambda} \notin \overline{D} \Rightarrow \lambda\left(\dfrac{1}{\lambda} - U^{-1}\right)$ is invertible that is, $U - \lambda$ is invertible.

$\Rightarrow \qquad \sigma(U) \subset T$

For fixed $\theta \in [0, 2\pi]$ and $n \in Z^+$

Let $x^n \in l_2(Z)$ be defined by

$$x_k^n = \begin{cases} (2^{n+1})^{-1/2} & e^{-ik\theta} \text{ for } |k| \le n \\ 0 & \text{Otherwise} \end{cases}$$

$x^n$ is a unit vector and

$$\lim_{n \to \infty} (U - e^{i\theta})x^n = 0$$

$\Rightarrow \qquad U - e^{i\theta}$ is not bounded below

$\Rightarrow \qquad e^{i\theta} \in \sigma(U)$

$\Rightarrow \qquad \sigma(U) = T$

**Definition 7.3.3**    $T \in B\ (H)$ is bounded below if $\exists \epsilon > 0$ such that $\|Tx\| \ge \epsilon \|x\|$ for each $x \in H$.

**Theorem 7.3.5**    $T$ is invertible iff $T$ is bounded below and has dense range.

**Proof.** $\Rightarrow\ T$ is invertible $\left\|\overset{-1}{T}\ Tx\right\| \le \left\|T^{-1}\right\|\ \|Tx\|$

So $\qquad \|Tx\| \ge \dfrac{1}{\left\|\overset{-1}{T}\right\|}\left\|\overset{-1}{T}\ Tx\right\| = \dfrac{1}{\left\|\overset{-1}{T}\right\|}\ \|x\|$

$\Rightarrow T$ is bounded below.

Conversely let $T$ be bounded below.

$\Rightarrow \qquad \exists \epsilon > 0 \qquad \|Tx\| \ge \epsilon \|x\|$

$\Rightarrow \quad \{Tx_n\}_n$ is a Cauchy sequence in Rng $T = H$.

$$\|x_n - x_m\| \le \frac{1}{\epsilon} \|T_{x_n} - T_{x_m}\|$$

$\Rightarrow \quad \{x_n\}$ is a Cauchy sequence.

Hence $\qquad x = \lim_{n\to\infty} x_n$

$\Rightarrow \qquad Tx = \lim_{n\to\infty} T_{x_n}.$

Since $T$ is bounded below, $T$ is 1–1

$\Rightarrow \quad T^{-1}$ is well defined.

If $y = Tx$, then

$$\left\| T^{-1} y \right\| = \|x\| \le \frac{1}{\epsilon} \|Tx\| = \frac{1}{\epsilon} \|y\|$$

$\Rightarrow \quad T^{-1}$ is bounded.

Spectrum of Bounded Self-adjoint Linear Operator

**Theorem 7.3.6** $\quad T:H \to H$ be self-adjoint.

Then $\lambda \in \rho(T)$ iff $\exists$ a $c > 0$ such that for all $x \in H$,

$$\|T_\lambda x\| \ge c\|x\|, \quad T_\lambda = T - \lambda I$$

**Proof.** $\lambda \in \rho(T) \Rightarrow R_\lambda = T_\lambda^{-1}$ exists and bounded.

Say $\|R_\lambda\| = k$, where $k > 0$ since $R_\lambda \ne 0$.

Now, $\qquad I = R_\lambda T_\lambda.$
For every $x \in H$,

$$\|x\| = \|R_\lambda T_\lambda x\| \le \|R_\lambda\| \|T_\lambda x\|$$

$$= k \|T_\lambda x\|$$

$\Rightarrow \qquad \|T_\lambda x\| \ge c \|x\|, \ c = \frac{1}{k}.$

Conversely, let the condition hold. We now show that

(i) $T_\lambda : H \to T_\lambda(H)$ is bijective,

(ii) $T_\lambda(H)$ is closed and dense in $H$.

For (i) suppose that $T_\lambda x_1 = T_\lambda x_2$. Since the condition holds for $c > 0$, we have

$$0 = \left\| T_\lambda x_1 - T_\lambda x_2 \right\| = \left\| T_\lambda(x_1 - x_2) \right\| \geq c \left\| x_1 - x_2 \right\|.$$

Hence $\left\| x_1 - x_2 \right\|$, thus $x_1 = x_2$.

(ii) Let $x_0 \in T_\lambda(H)$. Then for all $x \in H$ we have

$$0 = \langle T_\lambda x, x_0 \rangle = \langle Tx, x_0 \rangle - \lambda \langle x, x_0 \rangle$$

Since $T$ is self-adjoint, we have

$$\langle x, Tx_0 \rangle = \langle Tx, x_0 \rangle = \langle x, \bar{\lambda} x_0 \rangle.$$

So $Tx_0 = \bar{\lambda} x_0$. We claim that $x_0 = 0$. If $x_0 \neq 0$, then $\bar{\lambda}$ is an eigen-value of $T$ and since $T$ is self-adjoint, $\lambda$ is an eigen-value of $T$ and this contradicts that

$$0 = \left\| T_\lambda x_0 \right\| \geq c \left\| x_0 \right\| > 0.$$

Thus $T_\lambda(H) = \{0\}$, so $T_\lambda(H) = H$ and hence $T_\lambda(H)$ is dense in $H$.

For closedness of $T_\lambda(H)$, let $y \in T_\lambda(H)$. Then $\exists$ a sequence $\{y_n\}$ in $T_\lambda(H)$ so that $y_n \to y$. Since $y_n \in T_\lambda$, $y_n = T_\lambda(x_n)$ for $x_n \in (H)$. By the given condition

$$\left\| x_n - x_m \right\| \leq \frac{1}{c} \left\| T_\lambda(x_n - x_m) \right\| = \frac{1}{c} \left\| y_n - y_m \right\|$$

so $\{x_n\}$ is Cauchy in $H$ and hence $x_n \to x \in H$. Since $T$ is continuous, $T_\lambda$ is so. Thus

$$y_n = T_\lambda x_n \to T_\lambda x = y \in T_\lambda(H).$$

Hence $T_\lambda(H)$ is closed. Also $T_\lambda(H) = H$ Thus $R_\lambda = T_\lambda^{-1}$ is defined on $H$ and is bounded. Therefore $\lambda \in \rho(T)$.

**Theorem 7.3.7**   $T : H \to H$ be self-adjoint. Then $\sigma(T)$ is real.

**Proof.**   We show that $\lambda = \alpha + i\beta$, $\beta \neq 0$ must belong to $\rho(T)$. So that $\sigma(T) \subset R$.

For $$x \neq 0 \text{ in } H,$$

$$\langle T_\lambda x, x \rangle = \langle Tx, x \rangle - \lambda \langle x, x \rangle$$

$\langle x, x \rangle$ and $\langle Tx, x \rangle$ are real since $T$ is self-adjoint. So

$$\overline{\langle T_\lambda, x \rangle} = \langle Tx, x \rangle - \bar{\lambda} \langle x, x \rangle$$

$$\bar{\lambda} = \alpha - i\beta.$$ By subtraction

$$\overline{\langle T_\lambda x, \, x\rangle} - \langle T_\lambda x, \, x\rangle$$

$$= (\lambda - \bar{\lambda}) \langle x, \, x\rangle = 2i\beta \, \|x\|^2.$$

The left side is $-2i \, \mathrm{Im} \, \langle T_\lambda x, \, x\rangle$. The later cannot exceed the absolute value, so that dividing by 2, taking absolute value and applying Schwarz inequality, we obtain

$$|\beta| \|x\|^2 = |Im\langle T_\lambda x, \, x\rangle|$$

$$\leq |\langle T_\lambda x, \, x\rangle| \leq \|T_\lambda x\|\|x\|.$$

Dividing by $\|x\| \neq 0$ we get

$$|\beta|\|x\| \leq \|T_\lambda x\|$$

If $$\beta \neq 0, \, \lambda \in \rho \, (T)$$

So $\beta = 0$ and hence $\sigma(T) \subset R$.

**Theorem 7.3.8** $T:H \to H$ be self-adjoint.

$$\sigma(T) \subseteq [m_T, \, M_T]$$

where $$m_T = \inf_{\|x\|=1} \langle Tx, \, x\rangle, \, M_T = \sup_{\|x\|=1} \langle Tx, \, x\rangle$$

**Proof.** $\sigma \, (T) \subseteq R$

We show that any real $\lambda = M + c$ belongs to $\rho(T)$.

For every $x \neq 0$, and $v = \|x\|_x^{-1}$

We have $x = \|x\|v$ and

$$\langle Tx, \, x\rangle = \|x\|^2 \langle Tv, \, v\rangle$$

$$\leq \|x\|^2 \sup_{\bar{v}=1} \langle T\bar{v}, \, v\rangle$$

$$= \langle x, \, x\rangle M$$

Hence $\langle Tx, \, x\rangle \geq - \langle x, \, x\rangle M$

By Cauchy-Schwarz inequality we get

$$\|T_\lambda x\|\|x\| \geq - \langle T_\lambda x, \, x\rangle$$

$$= - \langle Tx, \, x\rangle + \lambda \langle x, \, x\rangle$$

$$\geq (- M + \lambda) \langle x, \, x\rangle = c\|x\|^2$$

where $c = \lambda - M > 0$ by assumption.

Dividing by $\|x\|$ we get

$\left\|T_{\lambda}x\right\| \geq c\left\|x\right\|$. Hence $\lambda \in \rho(T)$. For $\lambda < m$, the proof is similar.

## 7.4  NUMERICAL RANGE

If $T \in B(H)$ is self-adjoint we define the numerical range

$$W(T) = \{\langle Tx, x\rangle : \|x\| = 1\}$$

and numerical radius

$$W(T) = \sup\{\langle Tx, x\rangle : \|x\| = 1\}$$

**Theorem 7.4.1**  If $T$ is bounded linear self-adjoint,

then $$W(T) = \sup\{|\langle Tx, x\rangle| : \|x\| = 1\} = \|T\|$$

**Proof.**  We first prove that

$$W(T) \leq \|T\|$$

for any $T \in B(H)$.

We have for $\|x\| = 1$,

$$W(T) \leq |\langle Tx, x\rangle| \leq \|Tx\|\|x\| \leq \|T\|\|x\|^2 = \|T\|$$

We now show that

$$\|T\| \leq W(T).$$

Consider the identity

$$4\|Tx\|^2 = \langle T(\beta x + \beta^{-1}Tx), \beta x + \beta^{-1}Tx\rangle$$
$$- \langle T(\beta x - \beta^{-1}Tx), \beta x - \beta^{-1}Tx\rangle$$

This is true for any $x$ and any real $\beta \neq 0$.
It follows from the definition of $w(T)$:

$$4\|Tx\|^2 \leq W(T)\|\beta x + \beta^{-1}Tx\|^2$$
$$+ W(T)\|\beta x - \beta^{-1}Tx\|^2$$

By Parallelogram Law,

$$4\|Tx\|^2 \in 2W(T)(\beta^2\|x\|^2 + \beta^{-1}\|Tx\|^2)$$

Take $$\beta^2 = \frac{\|Tx\|}{\|x\|}.$$

This completes the proof.

We now give two examples of unbounded operators.

## 7.5　MULTIPLICATION OPERATORS

$$T: D\,(T) \to L^2\,(-\infty, +\infty) \qquad\qquad \dots(7.5)$$

$$x \to tx$$

where $D\,(T) \subset L^2\,(-\infty, +\infty)$

The domain $D\,(T)$ consists of all $x \in L^2\,(-\infty, +\infty)$ such that we have $Tx \in L^2$ $(-\infty, +\infty)$, that is,

$$\int_{-x}^{+x} t^2 |x(t)|^2\, dt < \infty. \qquad\qquad \dots(7.6)$$

This implies that $D\,(T) \neq L^2\,(-\infty, +\infty)$. For instance, an $x \in L^2\,(-\infty, +\infty)$ not satisfying (7.6) is given by

$$x(t) = \begin{cases} 1/t & \text{if } t \geq 1 \\ 0 & \text{if } t < 1 \end{cases};$$

hence $x \notin D\,(T)$.

Obviously, $D\,(T)$ contains all functions $x \in L^2\,(-\infty, +\infty)$ which are zero outside a compact interval. It can be shown that this set of functions is dense in $L^2\,(-\infty, +\infty)$. Hence $D\,(T)$ is dense in $L^2\,(-\infty, +\infty)$.

The multiplication operator $T$ defined by (7.5) is not bounded.

**Proof.** We take

$$x_n(t) = \begin{cases} 1 & \text{if } n \leq t < n+1 \\ 0 & \text{elsewhere} \end{cases}$$

Clearly, $\qquad \|x_n\| = 1$ and

$$\left\|Tx_n\right\|^2 = \int_{n}^{n+1} t^2\, dt > n^2.$$

This shows that $\left\|Tx_n\right\| / \left\|x_n\right\| > n$, where we can choose $n \in N$ as large as we please.

## 7.6　DIFFERENTIATION OPERATOR

$$\text{Dom } D \to L^2\,(-\infty, +\infty)$$

$$x \to ix' \qquad\qquad \dots(7.7)$$

where $x' = dx/dt$ and i helps to make $D$ self-adjoint. By definition, the domain of $D$ consists of all $x \in L^2\,(-\infty, +\infty)$ which are absolutely continuous on every compact interval on $R$ and such that $x' \in L^2\,(-\infty, +\infty)$.

Dom $D$ contains the sequence $(\hbar_n)$ involving the Hermite polynomials, and $(\hbar_n)$ is total in $L^2\,(-\infty, +\infty)$. Hence Dom $D$ is dense in $L^2\,(-\infty, +\infty)$.

The differentiation operator $D$ defined by (7.7) is unbounded.

**Proof.** $D$ is an extension of

$$D_0 = D|_y$$

where $Y = \mathrm{Dom}\, D \cap L^2\,[0,\,1]$ and $L^2\,[0,\,1]$ is regarded as a subspace of $L^2\,(-\infty,\,+\infty)$. Hence if $D_0$ is unbounded, so is $D$. We show that $D_0$ is unbounded.

Let

$$x_n(t) = \begin{cases} 1 - nt & \text{if } 0 \le t \le 1/n \\ 0 & \text{if } 1/n < t \le 1 \end{cases}.$$

The derivative is

$$x_{n'}(t) = \begin{cases} -n & \text{if } 0 < t < 1/n \\ 0 & \text{if } 1/n < t < 1 \end{cases}$$

We have

$$\|x_n\|^2 = \int_0^1 |x_n(t)|^2\, dt = \frac{1}{3n}$$

and

$$\|D_0 x_n\|^2 = \int_0^1 |xn'(t)|^2\, dt = n$$

and the quotient

$$\frac{\|D_0 x_n\|}{\|x_n\|} = n\sqrt{3} > n.$$

This shows that $D_0$ is unbounded.

## 7.7  COMPACT OPERATORS

**Definition 7.7.1**  An operator $T$ on a normed space is called compact if it maps every bounded set into a relatively compact set.

Observe that for linear operators, compactness is a stronger condition than boundedness; every compact operator is bounded, but converse is not true.

**Examples 7.7.1**  (i) Every bounded linear operator in a finite-dimensional space is compact.

(ii) In an infinite-dimensional space, any bounded linear operator whose range is finite-dimensional is compact.

(iii) In examples (1) and (3) the operators map every function into a linear combination of $\sin x$ and $\cos x$, therefore into a two-dimensional subspace of $L_2\,[0,\,2\pi]$. Hence, they are compact operators.

(iv) Let $K(x, y)$ be continuous on

$$[0, 1] \times [0, 1]. \text{ Define}$$

$$(Tf)(x) = \int_0^1 K(x, y) f(y)\, dy$$

It can be easily checked that if $E = \{f \in C[0, 1] : \|f\|_\infty \leq 1\}$, then $T(E)$ is uniformly bounded and equi-continuous. By applying the Arzela Ascoli theorem, we observe that $T(E)$ is relatively compact and, therefore, $T$ is compact.

**Example 7.7.2**  Let $K(x, y)$ be continuous for $a \leq x, y \leq b$, and $f(y, z)$ be continuous for $a \leq y \leq b$ and all $z$. Then the operator

$$T : C[a, b] \to C[a, b]$$

defined by

$$(Tu)(x) = \int_a^b K(x, y) f(y, u(y))dy$$

is compact.

**Example 7.7.3**  The identity operator is bounded but not compact in an infinite-dimensional space. For example, the unit ball is bounded but not compact, and the identity operator therefore maps a bounded set (the unit ball) into a non-compact set (the unit ball).

**Definition 7.7.2**  The rank of a linear operator is the dimension of its range. If the rank of a linear operator (i.e., the dimension of its range) is finite, then it is called a finite-rank operator.

**Theorem 7.7.1**  Let $X$ and $Y$ be normed spaces and $T : X \to Y$ a linear operator. Then $T$ is compact if and only if it maps every bounded sequence $(x_n)$ in $X$ onto a sequence $(Tx_n)$ in $Y$ which has a convergent subsequence.

**Proof.**  If $T$ is compact and $(x_n)$ is bounded, then the closure of $(Tx_n)$ in $Y$ is compact and this shows that $(Tx_n)$ contains a convergent subsequence.

Conversely, assume that every bounded sequence $(x_n)$ contains a subsequence $(x_{n_x})$ such that $(Tx_{n_x})$ converges in $Y$. Consider any bounded subset $B \subset X$, and let $(y_n)$ be any sequence in $T(B)$. Then $y_n = Tx_n$ for some $x_n \in B$, and $(x_n)$ is bounded since $B$ is bounded. By assumption, $(Tx_n)$ contains a convergent subsequence. Hence $\overline{T(B)}$ is compact because $(y_n)$ in $T(B)$ is arbitrary, this shows that $T$ is compact.

**Theorem 7.7.2**  Let $X$ and $Y$ be normed spaces and $T : X \to Y$ a linear operator. Then:

(a) If $T$ is bounded and $\dim T(X) < \leq \infty$, the operator $T$ is compact.

(b) If $\dim X < \infty$, the operator $T$ is compact.

**Proof.**  (a) Let $(x_n)$ be any bounded sequence in $X$. Then the inequality

$\|Tx_n\| \le \|T\| \, \|x_n\|$ shows that $(Tx_n)$ is bounded. Hence $(Tx_n)$ is relatively compact since $\dim T(X) < \infty$. It follows $(Tx_n)$ has a convergent subsequence. Since $(x_n)$ is an arbitrary bounded sequence in $X$, the operator $T$ is compact.

(b) Follows from (a) by noting that $\dim X < \infty$ implies boundedness of $T$ and $\dim T(X) \le \dim X$.

The following theorem states conditions under which the limit of a sequence of compact linear operators is compact. The theorem is also important as a tool for proving compactness of a given operator by exhibiting it as the uniform operator limit of a sequence of compact linear operators.

**Theorem 7.7.3**   Let $(T_n)$ be a sequence of compact linear operators from a normed space $X$ into a Banach space $Y$. If $(T_n)$ is uniformly operator convergent, say, $\|T_n - T\| \to 0$ then the limit operator $T$ is compact.

**Proof.**   We show that for any bounded sequence $(x_m)$ in $X$ the image $(Tx_m)$ has a convergent subsequence.

Since $T_1$ is compact, $(x_m)$ has a subsequence $(x_{1.m})$ such that $(T_1 x_{1.m})$ is Cauchy. Similarly, $(x_{1.m})$ has a subsequence $(x_{2.m})$ such that $(T_2 x_{2.m})$ is Cauchy. Continuing in this way we see that the $(y_m) = (x_{m.m})$ is a subsequence of $(x_m)$ such that for every fixed positive integer n the sequence $(T_n y_m)m \in N$ is Cauchy. $(x_m)$ is bounded, say $\|x_m\| \le c$ for all m. Hence $\|y_m\| \le c$ for all $m$. Let $\varepsilon > 0$. Since $T_m \to T$, there is an $n = p$ such that $\|T - T_p\| < \varepsilon/3c$. Since $(T_p y_m)_{m \in N}$ is Cauchy, there is an $N$ such that

$$\|T_p y_j - T_p y_k\| < \frac{\varepsilon}{3} \qquad\qquad (j,\, k > N).$$

Hence we obtain for $j,\, k > N$

$$\|Ty_j - Ty_k\| \le \| Ty_j - T_p y_j\| + \|T_p y_j - T_p y_k\| + \|T_p y_k - Ty_k\|$$

$$\le \|T - T_p\|\,\|y_j\| + \frac{\varepsilon}{3} + \|T_p - T\|\,\|y_k\|$$

$$< \frac{\varepsilon}{3c}\, c + \frac{\varepsilon}{3c} + \frac{\varepsilon}{3c}\, c = \varepsilon$$

This shows that $(Ty_m)$ is Cauchy and converges since $Y$ is complete. Remembering that $(y_m)$ is a subsequence of the arbitrary bounded sequence $(x_m)$, this implies compactness of the operator $T$.

Note that the present theorem becomes false if we replace uniform operator converges by strong operator convergence $\|T_n x - T_x\| \to 0$. This can be seen from $T_n : l^2 \to l^2$ defined by $T_n x = (\xi_1, \dots \xi_n,\, 0,0,\dots)$ where

$x = (\xi_j) \in l^2$. Since $T_n$ is linear and bounded, $T_n$ is compact. Clearly, $T_n x \to x = 1x$, but $I$ is not compact. The following example illustrates how the theorem can be used to prove compactness of an operator.

**Example 7.7.4** Prove that $T_n : l^2 \to l^2$ defined by $y = (\eta_j) = Tx$ where $n_j = \xi_j / j$ for $j = 1, 2, \ldots$ is a compact operator.

**Solution** $T$ is linear. If $x = (\xi_j) \in l^2$, then $y = (\eta_j) \in l^2$. Let $T_n : l^2 \to l^2$ be defined by

$$T_n x = \left( \xi 1, \frac{\xi 2}{2}, \frac{\xi 3}{3}, \ldots, \frac{\xi n}{n}, 0, 0, \ldots \right).$$

$T_n$ is linear and bounded, and is compact. Furthermore,

$$\|(T - T_n)x\|^2 = \sum_{j=n+1}^{\infty} |\eta_j|^2 = \sum_{j=n+1}^{\infty} \frac{1}{j^2} |\xi_j|^2$$

$$\leq \frac{1}{(n+1)^2} = \sum_{j=n+1}^{\infty} |\xi_j|^2 \leq \frac{\|x\|^2}{(n+1)^2}.$$

Taking the supremum over all $x$ of norm 1, we see that

$$\| T - T_n \| \leq \frac{1}{n+1}.$$

Hence $T_n \to T$, and $T$ is compact.

# Nonlinear Operators

8

In this chapter we discuss certain nonlinear operators, which are useful in the study of variational inequalities and complementarity problem.

**Definition 8.1.1**  Let $X$ be a real normed linear space and let $X^*$ be the dual space of $X$. Let the pairing between $x \in X$ and $x^* \in X^*$ be denoted by $(x^*, x)$. Let $T$ be a mapping of a subset $D(T)$ of $X$ into $X^*$. $T$ is said to be **monotone** if

$$(T_x - T_y, x - y) \geq 0 \quad \text{for all } x, y \in D(T),$$

and **strictly monotone** if T is monotone and the strict inequality holds whenever $x \neq y$. $T$ is called $\alpha$ - **monotone** if there is a continuous strictly increasing function $\alpha : [0, \infty) \to [0, \infty)$ with $\alpha(0) = 0$ and $\alpha(r) \to \infty$ as $r \to \infty$ such that

$$(T_x - T_y, x - y) \geq \|x - y\| \, \alpha(\|x - y\|)$$

for all $x, y$ in $D(T)$. $T$ is **strongly monotone** if $\alpha(r) = cr$ for some constant $c > 0$. $T$ is **coercive** on a subset $K$ of $D(T)$ if there exists a function $c : (0, \infty \to [-\infty, \infty]$ with

$$c(r) \to \infty \quad \text{as } r \to \infty$$

such that

$$(Tx, x) \geq \|x\| c(\|x\|) \quad \text{for all } x \in K.$$

Thus $T$ is coercive on $K$ if $K$ is bounded, while $T$ is coercive on an unbounded $K$ if and only if

$$\frac{(Tx, x)}{\|x\|} \to \infty \quad \text{as} \quad \|x\| \to \infty, \, x \in K$$

$T$ is **hemicontinuous** if $D(T)$ is convex and for any $x, y$ in $D(T)$, the map $t \to T(tx + (1 - t)y)$ of $[0, 1]$ to $X^*$ is continuous for the natural topology of $[0,1]$ and the weak topology of $X^*$.

**Examples**  (a) Let $f : R \to R$ be a monotonically increasing function. Then $f$ is a monotone operator.

(b) Let $H$ be a Hilbert space and $T : H \to H$ be a compact self-adjoint linear operator. Then $T$ is a monotone operator if all the eigen-values of $T$ are non-negative.

(c) Let $H$ be a Hilbert space. An operator $T : H \to H$ is said to be nonexpansive if $\|Tx - Ty\| \leq \|x{-}y\|$   for all $x, y \in H$.

If $T$ is nonexpansive, then $I - T$ is a monotone operator.

(d) Let $H$ be a Hilbert space and $C$ a closed convex subset of $H$. Let $Px$ denote the point of minimum distance of $C$ from $x$, that is,

$$Px = \{z \in C : \|z - x\| = \inf_{y \in C} \|y - x\|\}$$

Then $P$ is a monotone operator on $H$.

**Proof  of (d)**

By the definition we have

$$\|Px_1 - x_1\|^2 \leq \|Px_2 - x_1\|^2$$

and

$$\|Px_2 - x_2\|^2 \leq \|Px_1 - x_2\|^2$$

Expanding the norms in terms of inner - product we get

$$(Px_1 - Px_2,\ x_1) \geq \frac{1}{2}\,[\|Px_1\|^2 - \|Px_2\|^2]$$

$$\|Px_2 - Px_1,\ x_2) \geq \frac{1}{2}\,[\|Px_2\|^2 - \|Px_1\|2]$$

Combining these two we get the required result.

(e) Let $H$ be a Hilbert space. Then an operator $T : H \to H$ is said to be accretive if

$$\|x - y\| \leq \|Tx - Ty\|  \quad \text{for all } x, y \in H.$$

Then $T : H \to H$ is monotone iff $(I + \lambda\,T)$ is accretive for every $\lambda > 0$.

**Proof of (e)**

For $x, y\ \varepsilon\ H$, we have

$$\|(x - y) + \lambda\,(T_x - T_y)\|^2$$

$$= \|x - y\|^2 + 2\lambda\,(T_y,\ x - y) + \lambda^2 \|T_x - T_y\|^2$$

Now if $T$ is monotone, it clearly follows that $(I + \lambda T)$ is accretive. Conversely, if $(I + \lambda T)$ is accretive, then we have

$$2\lambda(T_x - T_y,\ x - y) + \lambda^2 \|T_x - T_y\|^2 \geq 0$$

for every $\lambda > 0$. Hence

$$(T_x - T_y, x - y) \geq -\lambda \, \|T_x - T_y\|^2 \text{ for } \lambda > 0$$

Letting $\lambda \to 0$ we observe that $T$ is monotone.

**Theorem 8.1.1**   If $T : D(T) \subset X \to X^*$ is $\alpha$ monotone, then it is strictly monotone (hence monotone) and coercive. In particular every strongly monotone operator is strictly monotone and coercive.

**Proof.**   It follows from the definition that every $\alpha$ monotone operator is strictly monotone. To prove coercivity, let $T$ be $\alpha$ monotone. Then there exists, a continuous strictly increasing function $\alpha : [0,\infty) \to [0, \infty)$ with $\alpha(0) = 0$ and $\alpha( r ) \to \infty$ as $r \to \infty$ such that

$$(T_x - T_0, x) \geq \|x\| \, \alpha(\|x\|)$$

i.e., $$(T_x, x) \geq (T_0, x) + \|x\| \, \alpha(\|x\|)$$

i.e., $$\frac{(T_x, x)}{\|x\|} \geq \frac{(T_0, x)}{\|x\|} + \alpha(\|x\|)$$

$$\to \infty \text{ as } \|x\| \to \infty$$

Therefore $T$ is coercive and this completes the proof.

**Definition 8.1.2**   Let $X$ be nls and let $X^*$ be its dual. A map $T : X \to X^*$ is said to be a duality map if for any $x \, \varepsilon \, X$.

(i) $$(T_x, x) = \|T_x\| \, \|x\| \text{ and (ii) } \|T_x\| = \|x\|$$

A duality map can be constructed in any nls in the following way : By Hahn-Banach theorem, for any $x \, \varepsilon \, X$, there exists at least one bounded linear functional $y_x \, \varepsilon \, X^*$ such that $\|y_x\| = 1$ and $(y_x, x) = \|x\|$. Taking one such functional $y_x$ and setting $T_x = \|x\| y_x$ and $T(-x) = -\|x\| y_x$, we get $\|T_x\| = \|x\|$ and $(T_x, x) = \|T_x\| \, \|x\|$

**Theorem 8.1.2**   In general a duality map $T : X \to X^*$ is multivalued. It is single-valued if $X^*$ is strictly convex.

**Proof.**   A duality map $T : X \to X^*$ is multi-valued, i.e., to each $x \, \varepsilon \, X$, there exists a set $u_x$:

$$U_x \subset V_x = \{v \, \varepsilon x^*(v, x) = \|x\|^2, \|v\| \leq \|x\|\}$$

The set $v_x$ is convex : for if $v_1, v_2 \, \varepsilon \, V_x$,

Then, $v = \lambda v_1 + (1 - \lambda)v_2 \, \varepsilon \, v_x, \ 0 \leq \lambda \leq 1$, since

$$(v, x) = (\lambda v_1 + (1 - \lambda) v_2, x) = \|x\|^2,$$

$$\| \lambda v_1 + (1 - \lambda) v_2 \| \leq \| x \|$$

By the definition of $v_x$, we have $v = 0$ if $x = 0$; if $x \neq 0$, it follows from

$$\| x \|^2 = (v, x) \leq \| v \| \, \| x \| \text{ that } \| x \| \leq \| v \|,$$

i.e. $\qquad\qquad v_x = u_x$ where

$$u_x = \{ u \, \varepsilon x^* : (u, x) = \| x \|^2, \| u \| = \| x \| \}$$

Hence $u_x$ is convex. Since $u_x \, \varepsilon \, s_r^* = \{ u \varepsilon x^* : \| u \| \leq \| x \| \}$

It follows that if $x^*$ is strictly convex, $u_x$ consists of only one point and thus any duality map is single valued.

**Theorem 8.1.3**   If $T : X \to X^*$ is a duality map, then it is monotone and coercive. If further $X$ is strictly convex, then $T$ is strictly monotone.

**Proof**   Let $T : X \to X^*$ be a duality map. Then it follows immediately from the definition that $T$ is coercive. Suppose that there exists a pair of points $x, y$ of $X$ such that

$$(T_x - T_y, \, x\text{-}y) < 0$$

Then we have

$$(T_x, \, x) + (T_y, \, y) - (T_y, \, x) - (T_x, \, y) < 0$$

Since, $\qquad\qquad | (T_y, \, x) | \leq \| T_y \| \, \| x \|$

and

$$| (Tx, \, y) | \leq \| T_x \| \, \| y \|$$

if follows from the condition (i) of the Definition of duality map that

$$0 > \| T_x \| \| x \| + \| T_y \| \, \| y \| - \| T_x \| \, \| \| - \| T_y \| \, \| x \|$$

$$= \| T_x \| - \| T_y \|)(\| x \| - \| y \|) \geq 0$$

By condition (ii) of the Definition of duality map we have

$$(\| T_x \| - \| T_y \|) (\| x \| - \| y \|) \leq 0$$

This contradiction shows that $T$ is monotone.

Now assume that $X$ is strictly convex. Suppose that $(T_x - T_y, \, x\text{-}y) = 0$. By the above inequality we have $\| x \| = \| y \|$. If $x \neq y$, (Since $X$ is strictly convex) it follows from the fact that it assumes its norm on $\dfrac{x}{\| x \|}$, that

$$(T_x - y) < \| T_x \| \, \| y \|$$

Similarly,

$$(T_y - x) < \| T_y \| \, \| x \|$$

Hence

$$0 = (T_x - T_y, x - y)$$

$$= \|T_x\|\|x\| \ \| + \|T_y\| \ \|y\| - (T_x, y) - (Ty, x)$$

$$> (\|T_x\| - \|T_y\|)(\|x\| - \|y\|) \geq 0$$

This is a contradiction and hence it follows that $T$ is strictly monotone.

**Theorem 8.1.4**  Let $X$ be a real Banach space and $F : X \to X^*$ be a nonlinear operator. If the Gateaux derivative $F'(x)$ exists for every $x \ \varepsilon \ X$ and is positive semidefinite, then $F$ is monotone.

**Proof.**  We have

$$(Fx_1 - Fx_2, x_1 - x_2)$$

$$= (F(x_2 + \theta(x_1 - x_2))(x_1 - x_2), (x_1 - x_2))$$

$$= (F(x)(x_1 - x_2), x_1 - x_2) \text{ where } x =_2 + \theta(x_1 - x_2) 0 < \theta < 1$$

Since $F'(x)$ is positive semi-definite, the result follows.

**Theorem 8.1.5**  Let $f$ be a proper convex function defined on $X$. If $f$ is differentiable, then $\nabla f$ is monotone.

**Proof.**  Let $x, y \ \varepsilon \ X$. Let the function $\phi$ be defined by

$$\Phi(\lambda) = f(x_1 + \lambda(x_2 - x_1))$$

Then

$$\Phi(\lambda) = (\nabla f(x_1 + \lambda(x_2 - x_1)), x_2 - x_1)$$

This gives

$$(\nabla f(x_2) - \nabla f(x_1), x_2 - x_1)$$

$$= \Phi'(1) - \Phi'(0) \geq 0$$

Hence $\nabla f$ *is* monotone.

**Theorem 8.1.6**  Let $f$ be a proper differentiable function defined on $X$. If $\nabla f$ is monotone, then $f$ is convex.

**Proof.**  We have

$$|\lambda f(x) + (1 - \lambda)f(y)| - f(\lambda x + (1 - \lambda)y)$$

$$= \lambda|f(x) - f(\lambda x + (1 - \lambda)y| + (1 - \lambda)[f(y) - f(\lambda x + (1 - \lambda y]$$

$$= \lambda(\nabla f(\lambda x + (1 - \lambda)y + t_1(1 - \lambda)(y - x)), (1 - \lambda(y - x))$$

$$+ (1 - \lambda)(\nabla f(\lambda x + (1 - \lambda)y + t_2 (1 - \lambda(x - y), \lambda(x - y))$$

for some $0 < t_1, t_2 < 1$, because of Lagrange theorem for functionals.

Rearranging the above equality we get

$$|\lambda f(x) + (1 - \lambda)\, f(y)| -f\, |\lambda x + (1 - \lambda)y|$$

$$= \frac{\lambda}{t_1 + t_2}\, [\, \nabla f(z_1) - \nabla f(z_2),\, z_1 - z_2)|$$

where

$$z_1 = \lambda x + (1 - \lambda)y + t_1(1 - \lambda)\,(y - x)$$

$$z_2 = \lambda x + (1 - \lambda)y + t_2(1 - \lambda)(y - x)$$

Now the required result follows from the fact that $\nabla f$ is monotone.

# Variational Inequalities

**9**

In this chapter we shall discuss some basic properties of variational inequalities. Before we state the definition we shall first discuss some examples where variational inequalities arise.

**Example 9.1.1**   Let $I\,[a, b] \subset R$. Let $f$ be a real-valued differentiable function defined on $I$. Suppose, we seek for the points $x \in I$ for which

$$f(x) = \min_{y \in I} f(y)$$

Then three cases will arise in this case:

(i) $a < x < y \Rightarrow f'(x) = 0$

(ii) $a = x \Rightarrow f'(x) \geq 0$

(iii) $x = b \Rightarrow f'(x) \leq 0$

All these three cases can be put together as a single inequality as follows:

$$f'(x)\,(y - x) \geq 0 \quad \text{for all } y \in I.$$

This is an example of variational inequality.

**Example 9.1.2**   Let $K$ be a closed convex set in $R^n$ and let $f \colon K \to R$ be differentiable. We characterize the points $x \in K$ for which

$$f(x) = \min_{y \in K} f(y).$$

If there exists $x \subset K$ which satisfies the above equation and if $F(x) = \text{grad}\, f(x)$, then $x$ is a solution of the following inequality

$$x \in K \colon \quad (Fx, y - x) \geq 0 \quad \text{for all } y \in K.$$

Conversely, if $f$ is differentiable and convex and if the above inequality is satisfied by $x$, then

$$f(x) = \min_{y \in K} f(y)$$

In order to prove this, let $y \in K$. For $0 \leq \lambda \leq 1$, put $z = (1 - \lambda)\, x + \lambda y$ $= x + \lambda\,(y - x)$. Since $K$ is convex, it follows that $z \in K$. Define a map $\Phi: [0, 1] \to R$ by

$$\Phi(\lambda) = f(x + \lambda(y - x)).$$

Now $\Phi$ attains its minimum at $\lambda = 0$.

Hence

$$\Phi'(o) = (\operatorname{grad} f,\, y - x) \geq 0 \quad \text{for all } y \in K,$$

thus $x \in K$ satisfies the inequality

$$(Fx,\, y - x) \geq 0 \quad \text{for all } y \in K.$$

To prove the converse part, observe that since $f$ is convex we have

$$f(y) \geq f(x) + (Fx,\, y - x) \qquad \text{for all } y \in K,$$

But $\qquad (Fx,\, y - x) \geq 0 \qquad\qquad\qquad\quad$ for all $y \in K,$

Hence $\qquad f(y) \geq f(x) \qquad\qquad\qquad\qquad\quad$ for all $y \in K,$

In otherwords,

$$f(x) = \min_{y \in K} f(y)$$

**Example 9.1.3**    Let $\Omega$ be a bounded open domain in $R^n$ with the boundary $T$.

In some problems of mechanics we seek a real-valued function $x \to u(x)$ which, in $\Omega$, satisfies the classical equation

$$-\Delta u - u = f,\, f \in \Omega,\, u = \sum_{i=1}^{n} \frac{\partial_2 u}{\partial x_i^2} \qquad \ldots(9.1)$$

with boundary conditions

$$u \geq 0,\, \frac{\partial u}{\partial v} \geq 0,\, u\,\frac{\partial u}{\partial v} = 0 \text{ on } \Gamma \qquad \ldots(9.2)$$

where $\dfrac{\partial}{\partial v}$ denotes differentiation along the outward normal to $\Gamma$. If we write

$$J(v) = \frac{1}{2}\, a(v,\, v) - (f,\, v)$$

where

$$a(u,\, v) = \sum_{i=1}^{n} \int_{\Omega} \frac{\partial u}{\partial x_i}\, \frac{\partial v}{\partial x_i}\, dx + \int_{\Omega} uv,\, dx$$

$$(f,\, v) = \int_{\Omega} fv\, dx$$

and if we introduce the closed convex set $K$ defined by

$$K = \{v: v \geq 0 \text{ on } \Gamma\},$$

then the problem given by (9.1) and (9.2) is equivalent to finding $u \in K$ such that

$$J(u) = \inf_{v \in K} J(v).$$

This admits a unique solution $u$ characterized by

$$u \in K, \ a(u, v - u) \geq (f, v - u) \quad \text{for all } v \in K.$$

This is called a variational inequality problem.

We shall now state the problem in the most general setting.

Let $X$ be a reflexive real Banach space and let $X^*$ be its dual.

Let $T$ be a monotone hemicontinuous mapping from $X$ into $X^*$ and let $K$ be a nonempty closed convex subset of the domain $D(T)$ of $T$. Then a variational inequality is stated as follows:

$$x \in K: \ (Tx, y - x) \geq 0 \quad \text{for all } y \in K \qquad \ldots(9.3)$$

Any $x \in X$ which satisfies (9.3) is called a solution of the variational inequality.

We write $S(T, K)$ to denote the set of all solutions of the variational inequality (9.3). We shall, infact, consider a more general inequality which is stated as follows:

For each given element $w_0 \in X^*$,

$$x \in K: \ (Tx - w_0, y - x) \geq 0 \quad \text{for all } y \in K \qquad \ldots(9.4)$$

Inequality (9.3) can also be written by replacing the subset $K$ of $X$ by an extended real-valued function defined on $X$. For any subset $K$ of $X$, let $\delta_k$, called the indicator function of $K$, be the function defined on $X$ by

$$\delta_k(y) = \begin{cases} 0 & \text{if } y \in K \\ \infty & \text{if } y \notin K \end{cases}$$

Then it is easy to verify that $x \in K$ is a solution of (9.3) if and only if

$$(Tx, y - x) \geq \delta_k(x) - \delta_k(y) \quad \text{for all } y \in K.$$

Therefore we consider, as a generalization of inequality (9.3), the inequalities of the form:

$$x \in X: \ (Tx, y - x) \geq f(x) - f(y) \quad \text{for all } y \in X, \qquad \ldots(9.5)$$

where $f$ is an arbitrary extended real-valued function defined on $X$.

Observe that if $f = 0$, then (9.5) reduces to the VI (3) and if $T = 0$, then we are in the framework of the calculus of variations where we minimize the extended real-valued functional $f$ i.e., we have

$$f(x) \le f(y), \quad \forall y \in X$$

or
$$f(x) = \min_{y \in X} f(y)$$

We shall first discuss in this section briefly the problem in the calculus of variation and next in the subsequent section we shall discuss the problem of variational inequality.

The problem in the calculus of variation is to find a function $x$ on $|t_1, t_2|$ that minimizes the functional

$$F = \int_{t_1}^{t_2} f[x(t), \dot{x}(t), t]\, dt$$

We assume that the function $f$ is continuous in $x$, $\dot{x}$ and $t$, and has continuous partial derivatives w.r.t. $x$ and $\dot{x}$. We seek a solution in the space $D\lfloor t_1, t_2 \rfloor$ and assume that the end points $x(t_1)$, $x(t_2)$ are fixed.

Starting with a given admissible vector $x$, we consider the vectors of the form $x + h$ that are admissible. The class of such vectors $h$ is called the class of admissible variations.

In the case of fixed end points it is clear that the class of admissible variations is the subspace of $D\lfloor t_1, t_2 \rfloor$, consisting of functions, which vanish at $t_1$ and $t_2$. The necessary condition for the extremum problem is that for all such $h$,

$$\delta F(x; h) = 0$$

(*) A set $\Omega$ on which an extremum (minimum or maximum) problem is defined is called admission set.

$$D\lfloor t_1, t_2 \rfloor = \left\{ x: (p_1, p_2] \to \not\in \quad \begin{array}{l} \text{which are continuous and} \\ \text{have continuous derivative} \end{array} \right\}$$

But

$$\delta F(x; b) = \frac{d}{da} \int_{t_1}^{t_2} f(x + \alpha h, x + \alpha h, t)\, dt \Big|_{\alpha} = 0$$

$$= \int_{t_1}^{t_2} fx(x, \dot{x}, t)\, h(t)\, dt + \int_{t_1}^{t_2} f_{\dot{x}}(x, \dot{x}, t)\, \dot{h}(t)\, dt,$$

and this differential is Frechet. Equate this differential to zero, assume that the functional $f_h$ has a continuous derivative w.r.t $t$ when the optimal solution is substituted for $x$, and integrate by parts to obtain

$$\delta F(x; h) = \int_{t_1}^{t_2} \left[ f_x(x, \dot{x}, t) - \frac{d}{dt} f_{\dot{x}}(x, \dot{x}, t) \right] h(t) \, dt + f_{\dot{x}}(x, \dot{x}, t) \, h(t) \Big|_{t_1}^{t_2} = 0$$

The boundary terms vanish for admissible $h$ and thus the necessary condition is

$$\int_{t_1}^{t_2} f_x(x, x, t) - \frac{d}{dt} f_{\dot{x}}(x, x, t) = h(t) \, dt = 0$$

for all $h \in D\,[t_1, t_2]$ vanishing at $t_1$, $t_2$.

Since $h(t)$ is continuous it follows:

that the extremal $x$ must satisfy the Euler-Lagrange equation

$$f_x(x, \dot{x}, t) - \frac{d}{dt} f_{\dot{x}}(x, \dot{x}, t) = 0.$$

We shall now consider some examples, which are of practical applications.

1.  Find the curve connecting two given points of minimum arc length.

Let the given points be $x(t_1)$, $x(t_2)$ corresponding to the real parameters $t_1, t_2$, we seek to minimize

$$\int_{t_1}^{t_2} \sqrt{1 + (\dot{x})^2} \, dt$$

Thus this is a special case of the problem : Minimize

$$\int_{t_1}^{t_2} f(x, \dot{x}, t) \, dt$$

where
$$f(x, \dot{x}, t) = \{1 + (\dot{x})^2\}^{\frac{1}{2}}$$

By Euler-Lagrange Equation we require

$$\frac{\partial f}{\partial x} - \frac{d}{dt} \frac{\partial f}{\partial \dot{x}} = 0$$

In this case
$$\frac{\partial f}{\partial x} = 0,$$

$$\frac{\partial f}{\partial \dot{x}} = \frac{1}{2} (1 + \dot{x}^2)^{-1/2} \cdot 2\dot{x} = \frac{\dot{x}}{\sqrt{1 + \dot{x}^2}}$$

Therefore

$$\frac{d}{dt} \left( \frac{\dot{x}}{\sqrt{1 + \dot{x}^2}} \right) = 0$$

or,
$$\frac{\dot{x}}{\sqrt{1 + \dot{x}^2}} = c, \quad \text{say}$$

or,
$$\dot{x} = c\,\sqrt{1 + \dot{x}^2}$$

or,
$$\dot{x}^2 = c^2\,(1 + \dot{x}^2)$$

or,
$$(1 - c^2)\,\dot{x}^2 = c^2$$

or,
$$\dot{x}^2 = \frac{c^2}{1 - c^2}$$

or,
$$\dot{x} = \frac{c}{\sqrt{1 - c^2}} = a, \quad \text{say.}$$

or,
$$\frac{dx}{dt} = a$$

or,
$$x = at + b$$

Thus the extremizing arc is the straight line connecting the two given points.

2.   Find the curve with given perimeter, which encloses maximum area. To be more specific, find the curve in $t - x$ plane having the end points $(-1, 0)$, $(1, 0)$; length $l$; and enclosing maximum area between itself and the $t$-axis.

Observe that we wish to maximize

$$\int_{-1}^{1} x(t)\, dt$$

subject to

$$\int_{-1}^{1} f\sqrt{\dot{x}^2 + 1}\, dt = 1$$

Therefore we seek a stationary point of

$$\int_{-1}^{1} f(x, \dot{x}, t)\, dt$$

where

$$f(x, \dot{x}, t) = x + \lambda\sqrt{\dot{x}^2 + 1}$$

Therefore
$$\frac{\partial f}{\partial x} = 1,$$

$$\frac{\partial f}{\partial \dot{x}} = \lambda\,\frac{\dot{x}}{\sqrt{\dot{x}^2}} + 1$$

By Euler-Lagrange Equation we require

$$\frac{\partial f}{\partial x} - \frac{d}{dt}\,\frac{\partial f}{\partial \dot{x}} = 0$$

or,
$$1 - \lambda\,\frac{d}{dt}\,\frac{\dot{x}}{\sqrt{1 + \dot{x}^2}} = 0$$

or,
$$\frac{d}{dt}\left(\frac{\dot{x}}{\sqrt{1 + \dot{x}^2}}\right) = 1/\lambda$$

or,
$$\frac{\dot{x}}{\sqrt{1 + \dot{x}^2}} = 1/\lambda\ t + c$$

or,
$$\dot{x}^2 = (1 + \dot{x}^2)\left(\frac{1}{\lambda}\,t + c\right)^2$$

or,
$$\dot{x}^2 = \left[1 - \left(\frac{t}{\lambda}\,t + c\right)^2\right] = \left(\frac{t}{\lambda} + c\right)^2$$

or,
$$\dot{x}^2 = \frac{\left(\frac{t}{\lambda} + c\right)^2}{1 - \left(\frac{t}{\lambda} + c\right)^2}$$

or,
$$\dot{x} = \frac{\frac{t}{\lambda} + c}{\sqrt{1 - \left(\frac{t}{\lambda} + c\right)^2}}$$

or,
$$x = \int \frac{\left(\frac{t}{\lambda} + c\right)}{1 - \left(\frac{t}{\lambda} + c\right)^2}\,dt$$

Put
$$1 - \left(\frac{t}{\lambda} + c\right)^2 = u^2$$

or,
$$-2\left(\frac{t}{\lambda} + c\right) \cdot \frac{1}{\lambda}\,dt = 2u\,du$$

Therefore
$$x = \int - \lambda\,\frac{u\,du}{u} = -\,\lambda u$$

Therefore
$$x - x_1 = -\,\lambda\left[1 - \left(\frac{t}{\lambda} + c\right)^2\right]^{1/2}$$

or,
$$(x - x_1)^2 = \lambda^2\left[1 - \left(\frac{t}{\lambda} + c\right)^2\right]$$

$$= \lambda^2 (t + c\,\lambda)^2$$

or,
$$(x - x_1)^2 + (t + c\lambda)^2 = \lambda$$

Thus the answer is a circle.

3. Find the extremum of

$$\int_{t_1}^{t_2} f(x, \dot{x}, t)\, dt$$

where
$$f(x, \dot{x}, t) = t\sqrt{1 + \dot{x}^2}$$

In this case
$$\frac{\partial f}{\partial x} = 0,$$

$$\frac{\partial f}{\partial \dot{x}} = \frac{t2\dot{x}}{2\sqrt{1 + \dot{x}^2}} = \frac{t\dot{x}}{\sqrt{1 + \dot{x}^2}}$$

By Euler Lagrange equation we require

$$\frac{\partial f}{\partial x} - \frac{d}{dt}\frac{\partial f}{\partial \dot{x}} = 0$$

or,
$$\frac{d}{dt}\left(\frac{t\dot{x}}{\sqrt{1 + \dot{x}^2}}\right) = 0$$

or,
$$\frac{t\dot{x}}{\sqrt{1 + \dot{x}^2}} = a$$

or,
$$t^2 \dot{x}^2 = a^2(1 + \dot{x}^2)$$

or,
$$(t^2 - a^2)\,\dot{x}^2 = a^2$$

or,
$$\dot{x} = \frac{a}{\sqrt{t^2 - a^2}}$$

or,
$$x = a \int \frac{dt}{\sqrt{t^2 - a^2}} + b$$

or,
$$= a \cosh^{-1}\frac{t}{a} + b$$

**Theorem 9.1.1**   Let $T$ be a monotone, hemi-continuous mapping of a subset $D\,(T)$ of $X$ into $X^*$ and $K$ a convex subset of $D(T)$. Then for a given element $w_0 \in X^*$, any solution of inequality (9.4) is also a solution of the inequality

$$(Ty - w_0,\ y - x) \geq 0\ \forall\ y \in K \qquad\qquad \ldots(9.6)$$

**Proof.**　Let $x \in K$ be a solution of (9.4). Since $T$ is monotone

$$(Ty - w_0, y - x) \geq (Tx, y - x) \geq (w_0, y - x)$$

for all $y \in K$. Hence $x$ is a solution of the inequality (9.6).

　　Conversely, let $x \in K$ be a solution of (6), and let $y$ be an arbitrary element of $K$. Since $K$ is convex, the vector $y_t = (1 - t) x + ty$, $0 < t < 1$, belongs to $K$. Hence by (9.6).

$$(Ty_t - w_0, y_t - x) \geq 0$$

i.e.　　　　　$$(Ty_t - w_0, y - x) \geq 0.$$

　　Since $T$ is a solution of (4) and this completes the proof.

**Theorem 9.1.2**　Let $T$ be a hemicontinuous map of $X$ into $X^*$. Suppose that for any pair of vectors $x_0 \in K$ and $w_0 \in X^*$,

$$(Ty - w_0, y - x_0) \geq 0 \quad \text{for all } y \geq X \qquad \qquad ...(9.7)$$

Then　　　　　$$Tx_0 = w_0.$$

**Proof.**　Suppose that $Tx_0 \neq w_0$. Then by the definition of the norm, there exists a nonzero vector $x \in X$ such that

$$(Tx_0 - w_0, z) > \frac{1}{2} \|z\| \, \|Tx_0 - w_0\| > 0 \qquad \qquad ...(9.8)$$

Since $T$ is hemicontinuous, it follows that for sufficiently small $t$,

$$|(T(x_0 - tz) - Tx_0, z)| \leq \frac{1}{3} \|z\| \, \|Tx_0 - w_0\| \qquad \qquad ...(9.9)$$

Now by (9.7) we may write

$$\left(T(x_0 - tz) - w_0, (x_0 - tz) - x_0\right) \geq 0.$$

giving

$$(T(x_0 - tz) - Tx_0, - z) + (Tx_0 - w_0 - tz) \geq 0$$

or

$$(T(x_0 - tz) - Tx_0, -z) \geq (Tx_0 - w_0, z)$$

Hence by (9.8) it follows that

$$(T(x_0 - tz) - Tx_0, z) > \frac{1}{2} \|z\| \, \|Tx_0 - w_0\|$$

　　This contradicts (9.9) and proves the result.

　　The following result gives uniqueness of solutions when it exists.

**Theorem 9.1.3**　If the mapping $T$ from $X$ into $X^*$ is strictly monotone, then the inequality (9.4) can have atmost one solution.

**Proof.**   Assume that $x_1$ and $x_2$ are two solutions for the inequality (9.4). Then for a given $w_0 \in X^*$ we have

$$(Tx_1 - w_0, x_2 - x_1) \geq 0$$

and

$$(Tx_2 - w_0, x_1 - x_2) \geq 0$$

Therefore we have

$$(Tx_1 - Tx_2, x_1 - x_2) \leq 0.$$

Since $T$ is strictly monotones, this is impossible unless $x_1 = x_2$ and this completes the proof.

**Theorem 9.1.4**   If either the mapping $T$ is strictly monotone or the function $f$ is strictly convex, then the inequality (9.5) can have atmost one solution.

**Proof.**   If $T$ is strictly monotone, the proof is similar to that of Theorem 9.1.3. Assume that $f$ is strictly convex. Suppose that $x_1$ and $x_2$ are two solutions of the inequality (9.5) and that $x_1 \neq x_2$. Then we have

$$(Tx_1, y - x_1) \geq f(x_1) - f(y)$$

and

$$(Tx_2, y - x_2) \geq f(x_2) - f(y)$$

for all $y \in X$. Putting $y = (x_1 + x_2)/2$ and adding we get

$$(Tx_1 - Tx_2, x_2 - x_1) \geq f(x_1) + f(x_2) - 2f\left(\frac{x_1 + x_2}{2}\right),$$

hence, since $T$ is monotone

$$2f\left(\frac{x_1 + x_2}{2}\right) \geq f(x_1) + f(x_2).$$

This contradicts the strict convexity of $f$ and completes the proof.

We shall now prove the following fundamental theorem for variational inequality.

**Theorem 9.1.5**   Let $T$ be a monotone hemicontinuous map of a closed convex subset $K$ of a reflexive real Banach space $X$, with $0 \in K$, into $X^*$ and if $K$ is not bounded, let $T$ be coercive on $K$. Then for each given element $w_0 \in K$ there is an $x \in K$ such that the inequality (9.4) holds, i.e.,

$$x \in K: (Tx - w_0, y - x) \geq 0 \quad \text{for all } y \in K.$$

We shall first establish some propositions. We need the following definition.

**Definition 9.1.1**  Let

$$c_r = \inf_{\|x\|=r} \frac{(Tx,\, x)}{\|x\|}$$

Note that by the hypothesis of Theorem (7), $c_r \to +\infty$ as $r \to +\infty$; we have

$$(Tx,\, x) \geq c(\|x\|)\, \|x\| \quad \text{for all } x \in K.$$

**Proposition 9.1.1**  There exists a constant $M$ which depends only on $c_r$ and on $\|w_0\|$ such that if $x_0$ is a solution of the inequality (4), then $\|x_0\| \leq M$.

***Proof.***  Let

$$(Tx_0 - w_0,\, y - x_0) \geq 0 \quad \text{for all } y \in K.$$

Since $0 \in K.$ we have

$$c(\|x_0\|)\, \|x_0\| \leq (Tx_0,\, x_0) \leq (Tx_0,\, w_0,\, x_0) + (w_0,\, x_0)$$

$$\leq \|w_0\|\, \|x_0\|$$

Hence $\qquad\qquad c(\|x_0\|) \leq \|w_0\|$

and $\qquad\qquad \|x_0\| \leq M(\|w_0\|,\, c_r).$

This completes the proof.

**Definition 9.1.2**  If $G \subset X \times X^*$, $g$ is said to be a monotone set if $[u,\, w]$, $[u_1,\, w_1] \in G$ implies that $(w - w_1,\, u - u_1) \geq 0$. $G$ is said to be maximal monotone if it is monotone and maximal in the monotone sets ordered by inclusion.

**Proposition 9.1.2**  Under the hypothesis of Theorem 9.1.5 suppose that $K$ has 0 as an interior point and let $G \in X \times X^*$ be given by

$$G = \left\{ \lfloor u,\, w \rfloor : u \in K,\, w = Tu + z \text{ where } (z,\, u - v) \geq 0 \text{ for all } v \in K \right\}.$$

Then $G$ is a maximal monotone set in $X \times X^*$.

***Proof.***  $G$ is monotone set since if $[u,\, w]$ and $[u_1,\, w_1] \in G$, with $w = Tu + z$, $w_1 = Tu_1 + z_1$, then $(w - w_1,\, u - u_1)$

$$= (Tu - Tu_1,\, u - u_1) + (z,\, u - u_1) + (z_1,\, u_1 - u) \geq 0$$

Suppose on the other hand that $[u_0,\, w_0] \in X \times X^*$ with

$$(w_0 - w,\, u_0 - u) \geq 0$$

for all $[u,\, w]$ in $G$. We assert first that $u_0 \in K$. Otherwise, there is some $s > 1$ such that $u_0 = sw_0$ for some $v_0$ on the boundary of $K$. Let $z_0 = 0$ be an element of $X^*$ such that $(z_0, v_0 - v) \geq 0$ for all $v \in K$. Since 0 is an interior point of $K$, $(z_0, v_0) > 0$. For each

$$\lambda > 0, \lfloor v_0, \lambda T v_0 + \lambda z_0 \rfloor \text{ lies in } G. \text{ Hence}$$

$$0 \leq (w_0 - T v_0 - \lambda z_0, u_0 - v_0) = (s - 1)(w_0 - T v_0 - \lambda z_0, v_0).$$

Cancelling $(s - 1) > 0$, we have

$$\lambda(z_0, z_0) \leq (w_0, v_0) - (T v_0, v_0)$$

which is a contradiction since $(z_0, v_0) > 0$ and $\lambda$ is arbitrary.

Hence $\qquad u_0 \in K$.

In addition, for each $u$ in $K$, $(u, Tu)$ lies in $G$. Hence

$$(Tu - w_0, u - u_0) \geq 0$$

Applying Theorem 1 we have

$$(Tu_0 - w_0, v - u_0) \geq 0 \quad \text{for all } v \in K$$

**Proposition 9.1.3**  Theorem 9.1.5 holds if $X$ is a finite dimensional Banach space $F$.

***Proof.***  We may suppose without loss of generally that $w_0 = 0$, that $F$ is a finite  dimensional Hilbert space with $F^* = F$, and that $K$ spans $F$ and hence has an interior point $v_0$ in $F$. Replacing $K$ by $K_0 = v_0 - K$ and defining a new map $T'$ on $K_0$ by $T'u = -T(v_0 - u)$, it is easy to verify that we may assume that 0 is an interior point of $K$ and the condition on $(Tu, u)$ is replaced by $(Tu, u - v_0) \geq c(\|u\|) \|u\|$ for a given $v_0 \in K$, with $c_r \to +\infty$ as $r \to +\infty$.

Let $G$ be the maximal monotone set in $F \times F^*$ constructed in Proposition 9.1.2. Then $nG$ is maximal monotone for each positive integer $n$. By theorem of Minty for each $n > 0$, there exists $[u_n, w_n] \in G$ such that $u_n + n w_r = 0$. Since $w_n = Tu_n + z_n$, where $(z_n, u_n - v) \geq 0$ for all $v \in K$ we have

$$- \left( \frac{1}{n} u_n, u_n - v_0 \right) = (w_n, u_n - v_0)$$

$$= (Tu, u_n - v_0) + (z_n, u_n - v_0)$$

$$\geq c(\|u_n\|) \|u_n\|$$

While $- \left( \dfrac{1}{n} u_n, u_n - v_0 \right) \leq \dfrac{1}{n} \|u_n\| \|v_0\|$.

Thus $c(\|u_n\|) \leq n^{-1} \|v_0\|$, and $\|u_n\| \leq M$, independent of $n$.

We may extract a subsequence which we again denote by $u_n$ such that $u_n \to u_0$ in $F$. Then $w_n \to 0$. For each $u \in K$,

$$(T_u - w_n, u - u_n) \geq 0$$

Taking the limit as $n \to \infty$ we have $(Tu, u - u_0) \geq 0$, $\forall u \in K$.

By Theorem 1, $(Tu_0, v - u_0) \geq 0$, $\forall v \in K$.

**Proof of Theorem 9.1.5**  It suffices to take $w_0 = 0$. For each finite dimensional subspace $F$ of $X$, let $K_F = K \cap F$, $j_F$ be the injection map of $F$ into $X$, $j_F^*$ the dual projection map of $X^*$ onto $F^*$. We set

$$T_F = j_F^* (T/K_F) : K_F \to F^*$$

Then $T_F$ satisfies the hypothesis of Proposition (9.1.3), and there exists $u_F$ in $K_F$ such that $(T_F u_F, u_F - v) = (Tu_F, u_F - v) \leq 0$, $v \in K_F$. By Proposition (9.1.1), since for $u \in K_F$, $(T_F, u, u) = (Tu, u) \geq c(\|u\|) \|u\|$ there exists a constant $M$ independent of $F$ such that $\|u_F\|$. Since $X$ is reflexive and $K$ is weakly closed, there exists $u_0 \in K$ such that for every finite dimensional $F$, $u_0$ lies in the weak closure of the set $V_F = \underset{F \subset F_1}{\cup} \{u_{F_1}\}$ Let $v$ be an arbitrary element of $K$, $F$ a finite dimensional subspace of $X$ which contains $v$. For $u_{F_1}$ in $v_F$, by Theorem (9.1.1), $(Tv, v - u_{F_1}) \geq 0$.

Since $(Tv, v - u_1)$ is weakly continuous in $v_1$, we have $(Tv, v - u_{F_1}) \geq 0$.

By Theorem (9.1.1), $(Tu_0, u_0 - v) \geq 0$ for all $v \in K$ and this completes the proof.

# Complementarity Problem

**10**

Several problems arising in various fields such as: mathematical programming, game theory, economics, mechanics and geometry have the mathematical formulation of a complementarity problem.

**Definition 10.1.1** Let $X$ be a reflexive real Banach space and let $X^*$ be its dual. Let $K$ be a closed convex cone in $X$ with $0 \in K$. The polar of $K$ is the cone $K^*$ defined by

$$K^* = \left\{ y \in X^*: (y, x) \geq 0 \ \forall x \in K \right\}$$

Obviously $K^* \neq \Phi$ since $0 \in K^*$. It is also easy to see that $K^*$ is a closed convex cone in $X^*$. Let $T$ be a map from $K$ into $X^*$. Then the complementarity problem (*CP* in short) is to find an $x \in X$ such that

$$x \in K, \ Tx \in K^*, \ (Tx, x) = 0$$

The following theorem proves the equivalence between the complementarity problem and the variational inequality over a closed convex cone. We write

$$S(T, K) = \{ x: x \in K, (Tx, y - x) \geq 0 \quad \text{for all } y \in K \}$$

and
$$C(T, K) = \left\{ x: x \in K, Tx, \in K^*, (Tx, x) = 0 \right\}$$

We have

**Theorem 10.1.1** $C(T, K) = S(T, K)$

**Proof.** It is obvious that $C(T, K) \subset S(T, K)$. To show the converse, let $x \in S (T, K)$. Take $y = 0 \in K$. Then we have $(Tx, x) \leq 0$. Taking $y = 2x$, we see that $(Tx, x) \geq 0$. Hence $(Tx, x) = 0$. It remains to show that $Tx \in K^*$. Assume the contrary that $Tx \notin K^*$. Then there exists $y_0 \in K$ such that $(Tx, y_0) < 0$. But since $x \in S (T, K)$ we have

$$0 > (Tx, y_0) \geq (Tx, x) = 0.$$

This contradiction completes the proof.

## Remarks 10.1.1

(a) It should be noted that the solution of a complementarity problem, if exists, is unique if the operator $T$ is strictly monotone. Since $C(T, K) = S(T, K)$ for a closed convex cone $K$, the proof is same as that of Theorem 9.1.3 of chapter 9.

(b) Regarding the existence it must be noted that the solution may not exist only under the assumption of hemicontinuity and monotonicity (even strict nonotonicity) of the operator $T$. For example, let $X = R$, $K = \{x \in R: x \geq 0\}$, so that $K = K^*$ and $K$ is a closed convex cone. Let $T: K \to R$ be defined by

$$Tx = -\frac{1}{1 + x}$$

Then $T$ is hemicontinuous and strictly monotone. $(Tx, x) = 0$ implies $x = 0$ but

$$T(0) = -1 \notin K^*.$$

We shall now discuss the existence of solutions of the complementarity problem. We have

**Theorem 10.1.2**  Let $T: K \to X^*$ be hemicontinuous, monotone and coercive. Then the complementarity problem (1) has a solution. In particular if $T$ is hemicontinuous and $\alpha$–monotone, then the solution exists and is unique.

**Proof.**  From Theorem 9.1.5 of previous chapter it follows that there exists an $x \in K$ such that

$$(Tx, y - x) \geq 0 \quad \text{for all } y \in K.$$

Now the result holds from Theorem 10.1.1. If $T$ is $\alpha$–monotone, then from Theorem 10.1.3 of previous chapter it follows that $T$ is monotone and coercive and hence the result holds.

**Theorem 10.1.3**  Let $T: K \to X^*$ be hemicontinuous, monotone and let $T_0 \in K^*$. Then the complementarity problem has a solution.

**Proof.**  For each $r \geq 0$ define

$$D_r = \{x \in K: \|x\| \leq r\}.$$

Observe that for each $r \geq 0$, $D_r$ is a nonempty closed convex and bounded set in $K$ with $0 \in D_r$. Therefore it follows from Theorem 9.1.5 of previous chapter that for each $r \geq 0$ there exists an $x_r \in D_r$ such that

$$(Tx_r, y - x_r) \geq 0 \quad \text{for all } y \in D_r$$

Since $0 \in D_r$ it follows that $(Tx_r, x_r) \leq 0$. Since $T$ is monotone we have

$$(Tx_r - T_0, x_r) \geq 0$$

and further if $T_0 \in K^*$ (or inparticular if $T_0 = 0$) we obtain

$$(Tx_r - x_r) \geq 0$$

Thus $(Tx_r, x_r) = 0$. Thus $x_r$ is a solution of the complementarity problems and this completes the proof.

**Theorem 10.1.4**  Let $T: C \to B^*$ be hemicontinuous and monotone such that there is an $x \in C$ with $Tx \in \text{int } C^*$. Then there is an $x_0$ such that

$$x_0 \in C, \ Tx_0 \in C \quad \text{and} \quad (Tx_0, x_0) = 0 \qquad \qquad \dots (10.1)$$

If further $T$ is strictly monotone, then there is a unique $x_0$ satisfying (10.1).

In order to prove the theorem we need the following result, which is due to Browder. See Browder (14).

**Proposition 10.1.1**  Let $T$ be a monotone hemicontinuous map of a closed, convex, bounded subset $K$ of $B$, with $0 \in K$, into $B^*$. Then there is an $x_0 \in K$ such that

$$(Tx_0, y - x_0) \geq 0 \quad \text{for all } y \in K.$$

## Proof of the Theorem

For any $e \in \text{int } C^*$ and each $r > 0$, $D_r(e)$ is clearly convex. Further the function $f: C \to R$ defined by $f(z) = (e, z)$ is obviously continuous. But $D_r(e) = f[0, r]$. Hence $D_r(e)$ is closed. Further $D_r(e)$ is bounded. Suppose to the contrary that it is not; then we can choose a sequence $\{z_n\}$ of isolated points in $D_r(e)$ satisfying.

$$\|z_n\| \to \infty \quad \text{as} \quad n \to \infty.$$

Let $y_n = \dfrac{z_n}{\|z_n\|}$; then $\|y_n\| = 1$ and $y_n \in D_r(e)$ for all $n$.

we therefore have a weakly convergent subsequence $y_{nk} \to y$. Since $D_r(e)$ is closed, $y_n \in D_r(e)$. Moreover

$$(e, y) = \lim_{k \to \infty} (e, y_{n_k}) = \lim_{k \to \infty} \left( e, \frac{z_{n_k}}{\|z_{n_k}\|} \right)$$

But

$$\left( e, \frac{z_{n_k}}{\|z_{n_k}\|} \right) = \frac{r}{\|z_{n_k}\|} \to 0 \quad \text{as} \quad k \to \infty.$$

Thus $(e, y) = 0$, Since $e \in \text{int } C^*$ we conclude that $y = 0$. This is a contradiction in view of the fact that $\|y_{n_k}\| = 1$ for every $k$.

Now for each $r > 0$, $D_r(e)$ is closed, convex and bounded. Therefore it follows from the above proposition that for each $r > 0$, there is an $x_r \in D_r(e)$ such that

$$(Tx_r, y - x_r) \geq 0 \qquad \text{for all } y \in D_r(e) \qquad \qquad \text{...(10.2)}$$

Since $0 \in D_r(e)$, it follows that $(Tx_r, x_r) \leq 0$. If there exist $e \in \text{int } C^*$ and $r > 0$ such that $x_r \in D_r^0(e)$, then there is some $\lambda > 1$ such that $\lambda x_r \in S_r(e) \subset D_r(e)$. Then we have that $(Tx_r, x_r) \leq (Tx_r, x_r) = \lambda(T_{x_r}, x_r)$. Since $(Tx_r, x_r) \leq 0$ it is impossible unless $(Tx_r, x) = 0$; thus $x_r$ satisfies (10.1). Now assume that $x_r \in S_r(e)$ for all $e \in \text{int } C^*$ and all $r > 0$. By the hypothesis there is an $x \in C$ with $Tx \in \text{int } C^*$. Set $e = Tx$. Choose $r > (Tx, x) \geq 0$. Now $x \in D_r^0(Tx)$ and since $T$ is monotone we have

$$(Tz, z - x) \geq (Tx, z - x) > 0 \quad \text{for all } z \in S_r(Tx) \qquad \qquad \text{...(10.3)}$$

But $x_r \in S_r(Tx)$ and hence $(Tx_r, x_r - x) > 0$.

Since $x \in D_r^0(Tx) \subset D_r(Tx)$, it follows from (10.2) that $(Tx_r, x - x_r) \geq 0$ and hence $(Tx_r, x_r - x) \leq 0$. Since this contradicts (10.3), the assumption "that $x_r \in S_r(e)$ for all $r$' has thus been shown not to hold when $e = Tx$. Thus the proof of the theorem is reduced to the previous case, the case where there exists $e \in \text{int } C^*$ and $r > 0$ such that $x \in D_r^0(e)$. If $T$ is strictly monotone it is easy to see that the solution is unique and this completes the proof of the theorem.

Now observe that if $e \in C^*$ but $e \in \text{int } C^*$, the sets $D_r(e)$ need not be bounded. In this case we cannot conclude $y = 0$ from the fact that $(e, y) = 0$. Consider the case when $B = R^2$, $C = R^2$ and $e = (1, 0)$. Then for each $r > 0$, $D_r(e)$ contains the positive $y$-axis and hence is unbounded.

We note that this theorem fails to hold if the requirement that there exists $x \in C$ with $Tx \in \text{int } C^*$ is dropped.

Take $B = R^3$, $C = \{(x, y, z) \in R^3 : x, z \geq 0, 2xz \geq y^2)\}$. Define $T$ by $T(x, y, z) = (x + 1, y + 1, 0)$. Then $T$ is monotone, hemicontinuous (even bounded). $(1, -1, 1) \in C$ and $T(1, -1, 1) = (2, 0, 0) \in C^*$. If $u = (x, y, z) \in C$ with $Tu \in C^*$, then $y = -1$ and hence $x > 0$. Hence for any such $u, (Tu, u) = x(x + 1) > 0$.


## 10.1  FINITE DIMENSIONAL CASE

Let $K$ be a closed convex cone in $R^n$ and $f$ a map from $K$ into $R^n$ such that

$$x \in K, \, f(x) \in K^*, \, (f(x), x) = 0$$

In particular, if $K = R_+^n$

$$\{x = (x_1, x_2,...,x_n) \in R^n : x_1 \le 0 \ i = 1, 2, \ldots, n\}$$

then the problem can be stated as follows:

$$x \ge 0, \ f(x) \ge 0, \ (f(x), x) = \sum_{i=1}^{n} x_i f_i(x) = 0.$$

If further $f(x) = Mx + b$ where $M$ is a given real square matrix of order $n$ and $b$ is a given column vector in $R^n$, then the above problem is called a linear complementary problem (LCP in short) and it can be stated as follows:

Find $w_1,\ldots,\ w_n$ and $x_1,\ldots,\ x_n$ such that

$$w = Mx + b, \ w \ge 0, \ x \ge 0, \ w_i \ x_i = 0, \ i = 1,\ldots, n.$$

Otherwise, in general, the problem is known as a nonlinear complementarity problem (NCP in short).

We shall now illustrate the LCP by a numerical example.

An example

As a specific example of an LCP in $R^n$, let

$$n = 2, \quad M = \begin{pmatrix} 2 & 1 \\ 1 & 2 \end{pmatrix}, \quad q = \begin{pmatrix} -5 \\ -6 \end{pmatrix}$$

In this case the problem is to solve:

$$w_1 - 2x_1 - x_2 = -5$$

$$w_2 - x_1 - 2x_2 = -6,$$

$$w_1, \ w_2, \ x_1, \ x_2 \ge 0, \ w_1 x_1 = w_2 x_2 = 0.$$

This can be expressed in the form of a vector equation as:

$$w_1 \begin{pmatrix} 1 \\ 0 \end{pmatrix} + w_2 \begin{pmatrix} 0 \\ 1 \end{pmatrix} + x_1 \begin{pmatrix} -2 \\ -1 \end{pmatrix} + x_2 \begin{pmatrix} -1 \\ -2 \end{pmatrix} = \begin{pmatrix} -5 \\ -6 \end{pmatrix} \qquad \ldots(10.4)$$

$$w_1, \ w_2, \ x_1, \ x_2 \ge 0, \ w_1 \ x_1 = w_2 x_2 = 0 \qquad \ldots(10.5)$$

Since $w_1 x_1 = 0$, in any solution of (10.5) at least one of the variables in the pair $(w_1, x_1)$ must be zero. A similar statement is true for the pair $(w_2, x_2)$, one way to solve this problem is to pick one variable from each of the pairs $(w_1, x_1)$, $(w_2, x_2)$ and to fix these variables at zero in the system of equations (10.4). The remaining variables in the system are usually called active variables. After eliminating the zero variables from (10.4) if the remaining system has a solution in which the active variables are nonnegative, that would provide a solution to (10.4) and (10.5). Denote the right-hand side constant vector in (10.4), $(-5, -6)$, as $(q_1, q_2)$. Pick the variables $w_1, w_2$ as zeros. Thus the active variables are $x_1, x_2$. By putting $w_1, w_2 = 0$ in (10.5) we get

$$x_1\begin{pmatrix} -2 \\ -1 \end{pmatrix} + x_2\begin{pmatrix} -1 \\ -2 \end{pmatrix} = \begin{pmatrix} -5 \\ -6 \end{pmatrix} = \begin{pmatrix} q_1 \\ q_2 \end{pmatrix} = q \qquad \ldots(10.6)$$

$$z_1 \geq 0,\ z_2 \geq 0$$

The system (10.6) has a solution iff the vector $q$ can be expressed as a nonnegative linear combination of the vectors $\begin{pmatrix} -2 \\ -1 \end{pmatrix}$ and $\begin{pmatrix} -1 \\ -2 \end{pmatrix}$. The set of all nonnegative linear combinations of $\begin{pmatrix} -2 \\ -1 \end{pmatrix}$ and $\begin{pmatrix} -1 \\ -2 \end{pmatrix}$ is a cone in the $q_1$, $q_2$ space.

The LCP has a solution in which the active variables are $x_1$, $x_2$ and $w_1 = w_2 = 0$ only if the given vector $q = \begin{pmatrix} -5 \\ -6 \end{pmatrix}$ lies in the above-mentioned cone. It is now evident that the point $(-5, -6)$ lies in the cone and the solution to (10.6) is $(x_1, x_2) = \left(\dfrac{4}{3}, \dfrac{7}{3}\right)$; hence the solution for the given LCP is $(w_1, w_2, x_1, x_2) = \left(0, 0, \dfrac{4}{3}, \dfrac{7}{3}\right)$.

As special cases we have the following results for $R^n$.

**Theorem 10.1.5**   Let $f: K \to R^n$ be continuous, monotone and coercive then NCP has a solution. In particular if $f$ is continuous and strongly monotone, then the solution exists uniquely.

**Theorem 10.1.6**   Let $f: K \to R^n$ be continuous, monotone and such that $f(0) \in K^*$ (or $f(0) = 0$). Then there exists a solution to the NCP.

**Theorem 10.1.7**   Let $f: K \to R^n$ be continuous, monotone and such that there exists an $x \in K$ with $f(x) \in$ int $K^*$. Then there exists a solution to the NCP. Lemke and Eaves discussed the existence of stationary points and the nature of the set of all stationary points of the pair $(f, K)$ in $R^n$ where $K = R^n_+$. Lemke discussed the linear case by considering affine function. A basic theorem of Lemke [30] states as follows.

**Theorem 10.1.8**   **(Lemke)**   Given an affine map $f: R^n_+$ and a $d \in R^n_+$ there is a piecewise affine map $x: R_+ \to R^n_+$ such that $x(t)$ is a stationary point of $(f, D^n_t)$ with

$$d.\ x(t) = t \quad \text{where}$$

$$D^n_t = \{x \in R^n_+ : d.\ x \leq t\}$$

**Definitions 10.1.2**   Let $M$ be square matrix of order $n$. $M$ is said to

$$y^T My \ = \sum_{i=1}^{n} \sum_{j=1}^{n} y_i \, m_{ij} \, y_j > 0 \quad \text{for all } 0 \neq y \in R^n,$$

Positive semi-definite if

$$y^T My \ \geq 0 \quad \text{for all } y \geq R^n,$$

Copositive matrix if

$y^T My \geq 0$ for all $y \geq 0$ and strict copositive if strict inequality holds for all $y \geq 0$,

Copositive plus if it is a copositive matrix and if

$$y^T (M + M^T) \ = 0, \text{ whenever } y \geq 0 \text{ satisfies } y^T My = 0,$$

*P*-matrix if all principal subdeterminents of $M$ are positive, *Q*-matrix if the LCP has a solution for every $q \in R^n$ degenerate matrix if all its principal subdeterminants are nonzero, degenerate matrix if it is not nondegenerate, $z$-matrix if $m_{ij} \leq 0$ for all $i \neq j$, *J*-matrix if

$$Mz \ \geq 0, \ z^T Mz \geq 0, \ z \geq 0 \Rightarrow z = 0.$$

# Appendix
# Semi-Inner-Product Space and Variational Inequality

I

In this chapter we discuss the concept of semi-inner product (sip.in short), which was introduced by Lumer in the year 1961 and subsequently studied by Giles and several other mathematicians. We then study variational inequality on sip.

Let $V$ be a complex vector space. A sip on $V$ is a complex function $[,]$ on $V \times V$ with the following properties: For $x, y, z, \in V$ and $\lambda \in \mathbb{C}$,

(i) $\lfloor x + y, z \rfloor = [x, z] + [y, z]$,

   $\lfloor \lambda x, y \rfloor = \lambda [x, y]$,

(ii) $[x, x] = 0$ for $x \neq 0$,

(iii) $|[x, y]|^2 \leq [x, x] [y, y]$

> $V$ along with a sip defined on it is called a sip space. A sip space $V$ has the homogeneity property when the sip satisfies

(iv) $[x, \lambda y] = \overline{\lambda} [x, y]$.

Recall that an inner-product is a complex function $(,)$ on $V \times V$ which has the following propertion: For $x, y, z \in V$ and $\lambda, \mu \in \mathbb{C}$,

(i) $(\lambda x + \mu y, z) = \lambda(x, z) + \mu(y, z)$,

(ii) $(x, x) \geq 0$, $(x, x) = 0$ iff $x = 0$,

(iii) $(x, y) = \overline{(y, x)}$

With the aim of carrying over Hilbert space type arguments to the theory of Banach spaces Lumer (7) introduced the concept of sip. But the generality of the axiom system defining the sip is a serious limitation of any extensive development of a theory of sip spaces parallel to the theory of inner-product spaces. Let $X$ be a normed linear space and let $X^*$ be its dual.

The unit ball of $X$ is

$$U = \{ x \in X : \|x\| \leq 1 \} \quad \text{and its boundary}$$

$$S = \{x \in X: \|x\| = 1\} \quad \text{is the unit sphere of } X.$$

$$U^* = \{f \in X^*: \|f\| \leq 1\} \quad \text{is the unit ball and}$$

$$S^* = \{f \in X: \|f\| = 1\} \quad \text{is the unit sphere of } X^*$$

The conjugate norm will also be denoted by $\|.\|$.

**Theorem AI.1.1** A sip space $V$ is a normed linear space with the norm $\|x\| = [x, x]^{1/2}$. Every normed linear space can be made into a sip space (in general, in infinitely many different ways) with the homogeneity property.

**Proof.** We first show that $\|x\| = [x, x]^{1/2}$ is a norm. We have

$$\|x + y\|^2 = [x + y, x + y] = [x, x + y] + [y, x + y]$$

$$\leq (\|x\| + \|y\|)\,(x + y)$$

Therefore

$$\|x + y\| \leq \|x\| + \|y\|.$$

Again

$$\|\lambda x\|^2 = [\lambda x, \lambda x] = \lambda[x, \lambda x] = |\lambda|\,\|x\|\,\|\lambda x\|$$

Therefore $\|\lambda x\| \leq |\lambda|\,\|x\|$. If $\lambda = 0$, then clearly equality holds. If $\lambda \neq 0$, then

$$\|x\| = \left\|\frac{1}{\lambda}\lambda x\right\| \leq \frac{1}{|\lambda|}\|\lambda x\|; \text{ and hence } \|\lambda x\| \geq |\lambda|\,\|x\|.$$

Therefore $\|\lambda x\| = |\lambda|\,\|x\|$.

On the other hand let $X$ be a normed linear space. For each $x \in X$, there exists by Hahn-Banach theorem at least one (and we choose exactly one) functional $W_X \in X^*$ such that $(x, W_x) = \|x\|^2$. Given any such mapping $W$ from $X$ into $X^*$ (and there exists in general for a given $X$ infinitely many such mappings), it is at once verified that $[x, y] = (x, W_y)$ defines a sip. Again for each $x \in S$, there exists by Hahn-Banach theorem atleast one (and we choose exactly one) $f_x \in S^*$ such that $f_x(x) = 1$. For $\lambda x \in V$, where $x \in S$ and $\lambda \in \mathbb{C}$, we choose $f_{\lambda x} \in V^*$ such that $f_{\lambda x} = \overline{\lambda}f$. Given one such mapping from $V$ into $V^*$, there exists in general infinite number of such mappings and it is readily verified that the function $[x, y] = f_y(x)$ satisfies the properties of sip including the homogeneity property.

This completes the proof.

**Theorem AI.1.2** A Hilbert space $H$ can be made into a sip space in a unique way. A sip space is an ip space if and only if the norm it induces satisfies parallelogram law.

**Proof.**  Let [,] be a sip on a Hilbert space $H$. Then $[x, y]$ is, for fixed $y \neq 0$, a bounded linear functional on $H$. Therefore by Riesz-Representation theorem there exists a $z \in H$ such that $[x, y] = (x, z)$, where the later bracket denotes the usual inner product. From this $\|y\| = \|z\|$, and from $\|y\|^2 = (y, z)$ by the strict Schwarz inequality it follows that $z = \lambda y$; but again $(y, \lambda y)$ so that $z = y$.

## AI.1  CONTINUOUS AND UNIFORM SIP SPACES

A continuous sip space is a sip space V where the sip has the additional property:

(v) For all $(x, y) \in S \times S$

$$\text{Re} \ \lfloor y, x + \lambda y \rfloor \ \rightarrow \ \text{Re} \ \lfloor x, y \rfloor \quad \text{for all } \lambda \rightarrow 0.$$

The space is a uniformly continuous sip space when the above limit is approached uniformly for all $(x, y) \in S \times S$.

A uniform sip space is a **uniformly continuous** sip space where the induced normed linear space is uniformly continuous and complete.

### Examples ($L_p$ space for $1 < p < \infty$)

The real Banach space $L_p(X, S, \mu)$, where $1 < p < \infty$, can readily be expressed as a uniform sip space with sip defined by

$$[y, x] \ = \ \frac{1}{\|x\|^{p-2}} \int_X y \, |x|^{p-1} \, \text{sgn} \, x \, d\mu$$

We first consider the case $2 \leq p < \infty$. It is obvious that the function satisfies the properties (i), (ii), (iv) of sip. We proceed to establish (iii).

By Holder's inequality we have

$$\left| \int_X y |x|^{p-1} \, \text{sgn} \, x \, d\mu \right|$$

$$\leq \left( \int_X y(x)p \, d\mu \right)^{1/q} \left( \int_X |y|^p \, d\mu \right)^{1/p}$$

where $\dfrac{1}{p} + \dfrac{1}{q} = 1$, and the required inequality follows.

Considering $x, y, z \in S$ we have

$$|[z, x] - [z, y]|$$

$$\leq \int_X \left| \left( |x|p^{-2} \, x - |y|^{p-2} \, y \right) z \right| \, d\mu.$$

If we assume $x(t) > 0$ and $|x(t)| > |y(t)|$, then

$$0 \leq |x(t)|^{p-2}\, x(t) - |y(t)|^{p-2}\, y(t)$$

$$\leq (p-1)\, |x(t)^{p-2}\, (x(t) - y(t))$$

In general

$$\left| |x|^{p-2}\, x - |y|^{p-2}\, y \right|$$

$$\leq (p-1)\, |x - y| \left( |x|^{p-2} + |y|^{p-2} \right)$$

Now by Holder's inequality, since $p \geq 2$,

$$\int_X |x|^{p-2}\, |(x-y)\, z|\, d\mu.$$

$$\leq \left( \int_X |(x-y)\, z|^{p/2}\, d\mu \right)^{1/p}$$

$$\leq \left( \int_X |x-y|^p\, d\mu \right)^{1/p}$$

Therefore

$$\left| [z, x] - [z, y] \right| \leq 2(p-1)\, \|x - y\|_p,$$

which implies that (v) holds uniformly for all $(x, y) \in S \times S$.

It has been established that such space are uniformly convex.

We now consider $L_p$ spaces where $1 < p \leq 2$. The representation theorem for continuous linear functionals $f_x$ corresponding to vector $x \in L_p$ is given by

$$f_x(z) = \frac{1}{\|x\|_p^{p-2}} \int_X y|x|^{p-1}\, \operatorname{sgn} x\, d\mu,\ \forall y$$

Further we have that the dual $L_p^*$ of an $L_p$ space $(2 \leq p < \infty)$ is itself a sip space defined by

$$\lfloor f_x, f_y \rfloor = 1/\|x\|_p^{p-2} \int_X y|x|^{p-1}\, \operatorname{sign} x\, d\mu$$

$$= \frac{1}{\|\eta\|_q^{q-2}} \int_X \xi|\eta|^{q-1}\, \operatorname{sgn} x\, d\mu$$

where

$$\frac{1}{p} + \frac{1}{q} = 1$$

$$\xi = \left( \frac{1}{\|x\|_p^{p-2}} \right) |x|^{p-1}\, \operatorname{sgn} x$$

$$\eta = \left( \frac{1}{\|y\|_p^{p-2}} \right) |y|^{p-1}\, \operatorname{sgn} y$$

From this sip for $L_p^*$ we deduce that $L_p^*$ where $(2 \le p < \infty)$ is $L_q$ where $\dfrac{1}{p} + \dfrac{1}{q}$ $= 1$. We conclude that $L_p$ space for $1 < p < \infty$ can be expressed as a uniform sip space with our defined sip. For $x$, $y$ in any sip space $V$, $x$ is said to be normal to $y$ and $y$ is transversal to $x$ if $[y, x] = 0$. A vector $x \in V$ is normal to a subspace $N$ and $N$ is transversal to $x$, if $x$ is normal to each $y \in N$.

A Banach space $X$ is said to be smooth at a point $x \in S$ if and only if there exists a unique hyperplane of support at $x$, that is, there exists only one continuous linear functional $l_x \in E^*$ with $\|l_x\| = 1$ and $l_x(x) = 1$. $X$ is said to be a smooth Banach space if it is smooth at every $x \in S$.

The norm of $X$ is said to be Gateaux differentiable if for all $x$, $y \in S$ and real $\lambda$.

$$\lim_{\lambda \to 0} \frac{\|x + \lambda y\| - \|x\|}{\lambda} \text{ exists.}$$

The norm is said to be uniformly Frechet differentiable if this limit is approached uniformly for $(x, y) \in S \times S$. Note that $X$ is smooth at $x \in S$ if and only if the norm is Gateaux differentiable at $x$. We have

**Theorem AI.1.3**　In a continuous sip space $x$ is normal to $y$ if and only if $\|x + \lambda y\| > \|x\|$ for all complex numbers $\lambda$.

**Proof.**　If $x$ is normal to $y$ then

$$\|x + \lambda y\|\, \|x\| \ge \big|[x + \lambda y, x]\big|$$

$$= \big|\|x\|^2 + \lambda[y, x]\big| = \|x\|^2$$

Hence

$$\|x + \lambda y\| \ge \|x\| \quad \text{for all complex } \lambda.$$

Conversely, if

$$\|x + \lambda y\|^2 - \|x\|\, \|x + \lambda y\| \ge 0$$

Hence

$$\operatorname{Re}[x, x + \lambda y] + \operatorname{Re} \lambda[y, x + \lambda y]$$

$$- \big|[x, x + \lambda y]\big| \ge 0$$

$$\Rightarrow \quad \operatorname{Re} \lambda[y, x + \lambda y] \ge 0 \quad \text{for all } \lambda \in \mathbb{C}$$

Hence for real $\lambda$,

$$\operatorname{Re}[y, x + \lambda y] \ge 0 \quad \text{for all } \lambda \ge 0$$

$$\le 0 \quad \text{for all } \lambda \le 0.$$

But by the continuity condition we have for real $\lambda$,

$$\mathrm{Re}[y,\, x + \lambda y] \qquad \rightarrow \qquad \mathrm{Re}\,[y,\, x]$$

through positive values for $\lambda \to 0+$ and through negative values for $\lambda \to 0-$.

Thus $\mathrm{Re}\,[y,\, x] = 0$.

For imaginary $\lambda$, say $\lambda = i\lambda_1$, $\lambda_1$ real,

$$\mathrm{Re}\,\lambda[y,\, x + \lambda y] = \lambda_1\,\mathrm{Re}[iy,\, x + \lambda_1 iy]$$

and again by the continuity condition

$$\mathrm{Re}\,[iy,\, x] = 0, \text{ i.e.}$$

$$[y,\, x] = 0$$

This completes the proof.

**Theorem AI.1.4**   A sip space is continuous (uniformly continuous) sip space iff the norm is Gateaux (uniformly Frechet) differentiable.

**Proof.**   Let $V$ be a sip space. For $x,\, y \in S$ and real $\lambda > 0$,

$$
\text{(i)} \qquad \frac{\|x + \lambda y\| - \|x\|}{\lambda} \geq \frac{\big|[x + \lambda y,\, x]\big| - \|x\|^2}{\lambda\|x\|}
$$

$$
\geq \frac{\mathrm{Re}[x + \lambda y,\, x] - \|x\|^2}{\lambda\|x\|}
$$

$$
= \frac{\mathrm{Re}\lfloor y,\, x\rfloor}{\|x\|}
$$

But also

$$
\text{(ii)} \qquad \frac{\|x + \lambda y\| - \|x\|}{\lambda} \leq \frac{\|x + \lambda y\|^2 - \big|[x,\, x + \lambda y]\big|}{\lambda\|x + \lambda y\|}
$$

$$
\leq \frac{\big|[x,\, x + \lambda y]\big| + \lambda\,\mathrm{Re}[y,\, x + \lambda y] - \big|[x,\, x + \lambda y]\big|}{\lambda\|x + \lambda y\|}
$$

$$
= \frac{\mathrm{Re}[y,\, x + \lambda y]}{\|x + \lambda y\|}
$$

(i) and (ii) show that the continuity (uniform continuity) property $\Rightarrow$ the norm is Gateaux (uniformly Frechet) differentiable and that the differential is $\dfrac{\mathrm{Re}[y,\, x]}{\|x\|}$.

Conversely, suppose the sip space has Gateaux (uniformly Frechet) differentiable norm. Since

$$\frac{\|x + \lambda y\| - \|x\|}{\lambda} \leq \frac{\mathrm{Re}\lfloor y, x\rfloor}{\|x\|}, \quad \lambda > 0$$

$$\geq \frac{\|x + \lambda y\| - \|x\|}{\lambda}, \quad \lambda < 0$$

Therefore

$$\lim_{\lambda \to 0} \frac{\|x + \lambda y\| - \|x\|}{\lambda} = \frac{\mathrm{Re}[y, x]}{\|x\|}.$$

For $x, y \in S$ and real $\lambda > 0$,

$$\frac{\|x + \lambda y\| - \|x\|}{\lambda} = \frac{\mathrm{Re}[x, x + \lambda y] + \lambda\,\mathrm{Re}[y, x + \lambda y] - \|x\|\,\|x + \lambda y\|}{\lambda \|x + \lambda y\|}$$

$$\geq \frac{\mathrm{Re}\,[y, x]}{\|x\|} \tag{iii}$$

From (i)

$$\lim_{\lambda \to 0^+} \inf\{\mathrm{Re}[x, x + \lambda y] - \|x\|\,\|x + \lambda y\|\} \geq 0$$

Since the norm is differentiable, we have from (iii) that

$$\lim_{\lambda \to 0^+} \inf \frac{\mathrm{Re}[x, x + \lambda y] - \|x\|\,\|x + \lambda y\|}{\lambda \|x + \lambda y\|}$$

is finite and $\geq 0$.

Hence

$$\lim_{\lambda \to 0^+} \frac{\|x + \lambda y\| - \|x\|}{\lambda}$$

$$\geq \lim_{\lambda \to 0^+} \sup \frac{\mathrm{Re}[y, x + \lambda y]}{\|x + \lambda y\|}$$

$$\geq \mathrm{Re}\lfloor y, x\rfloor\,\|x\|$$

and from (i) and (ii)

$$\lim_{\lambda \to 0^+} \frac{\mathrm{Re}[y, x + \lambda y]}{\|x + \lambda y\|} \quad \text{exists and is equal to} \quad \frac{\mathrm{Re}[y, x]}{\|x\|}$$

We now conclude that

$$\mathrm{Re}[y, x + \lambda y] \to \mathrm{Re}[y, x] \quad \text{as } \lambda \to 0$$

when norm is Gateaux differentiable and uniformly for $(x, y) \in S \times S$ when the norm is uniformly Frechet differentiable. This proves the theorem.

**Lemma AI.1.1**  In a continuous sip space which is uniformly convex and complete in its norm, there exists a nonzero vector normal to every proper closed vector subspace.

**Proof.**  It is well known that in uniformly convex Banach space, given a proper closed vector subspace $N$ and a vector $y \notin N$; there exists a unique nonzero vector $x_0 \in N$ such that

$$\|y - y_0\| = \inf \{ \|y - x\| : x \in N \}$$

Writing $z_0 = y - x_0$, we get

$$\|z_0\| \leq \|z_0 + x\| \quad \text{for all } x \in N,$$

i.e. $z_0$ is normal to $N$.

**Lemma AI.1.2**  A sip space is strictly convex if whenever $[x, y] = \|x\| \, \|y\|$, $x$, $y \neq 0$, then $y = \lambda x$ for some real $\lambda > 0$.

**Theorem AI.1.5**  (**Generalized Riesz – Fischer Theorem**)

In a continuous sip space $V$ which is uniformly convex and complete in its norm, to every continuous linear functional $f \in V^*$, there exists a unique vector $y \in V$ such that

$$f(x) = [x, y], \; x \in V$$

**Proof.**  (Existence). If $f(x) = 0$ for all $x \in V$, then we choose $y$ as the null vector in $V$.

If $f(x) \neq 0$ for some $x \in V$, the null space $N$ of $f$ is a proper closed vector subspace of $V$. Hence by Lemma 1 there exists a nonzero vector $y_0$ normal to $N$.

When $x \neq y_0$, $f(x) = [x, y] = 0$ for $y = \alpha \, y_0$ for any complex $\alpha$.

When $x = y_0$, $f(x) = [x, y] = f(y_0)$ for $y = \left( \overline{(f(y_0)}/\left(\|y_0\|^2\right)\right) y_0$.

Since each $x \in N$ can be represented in the form $x = z + \lambda y_0$ where $z \in N$, $y_0$ is normal to $N$ and

$$\lambda = \frac{f(x)}{f(y_0)}, \text{ we have for all } x \in V,$$

$$f(x) = f(z + \lambda y_0) = f(z) + \lambda f(y_0)$$

$$= [z, y] + \lambda [y_0, y]$$

$$= [z + \lambda y_0, y] = [x, y].$$

(Uniqueness). Suppose there exists vectors $y, y' \in V$, $y \neq y'$ such that

$$f(x) = \lfloor x, y \rfloor = \lfloor x, y' \rfloor \ \forall x \in \Lambda.$$

Then $\qquad\qquad [y, y] = [y, y'] \leq \|y\| \ \|y'\|,$

So $\qquad\qquad\qquad \|y\| = \|y'\| \text{ and hence } \|y\| = \|y'\|$

From $\|y\|^2 = [y, y']$ it follows that $\|y\| \ \|y'\|$ and from Lemma 2 that $y = y'$. This completes the proof.

**Theorem AI.1.6**  For a uniform sip space $M$, the dual space $M^*$ is a uniform sip space w. r. t. the sip defined by

$$\lfloor f_x, f_y \rfloor = \lfloor y, x \rfloor$$

**Proof.** $\qquad [f_z, f_x] + [f_z, f_y]$

$$= [x, z] + [y, z] = [x + y, z]$$

$$= [f_z, f_{x+y}].$$

It is known that a Banach space is uniformly Frechet differentiable iff it is uniformly convex. It follows that $M^*$ is uniformly convex. It is also well known that the dual of a Banach space is a Banach space.

**Theorem AI.1.7**  Every finite dimensional strictly convex, continuous sip space is a uniform sip space.

**Proof.**  In a continuous sip space the function $f$ defined on $S \times S$ by

$$f(x, y) = \text{Re}[y, x]$$

is continuous in the sense that

$$\text{Re}[y, x + \lambda y] \to \text{Re}[y, x] \quad \text{for all real } \lambda \to 0.$$

But in a finite-dimensional space, $S$ is compact and hence so is $S \times S$. Hence $\text{Re}[y, x + \lambda y] \to \text{Re}[y, x]$ for all real $\lambda \to 0$, uniformly for $(x, y) \in S \times S$. It is well known that every strictly convex finite-dimensional nls is uniformly convex and that every finite dimensional nls is complete.

This completes the proof.

**Theorem AI.1.8**  Let $X$ be a continuous sips which is uniformly convex and complete in its norm. If $A$ is a bounded linear operator from $X$ into itself, then there is a unique bounded linear operator $A^+$ such that

$$[Ax, y] = [x, A^+ y].$$

$A^+$ is called the generalized adjoint of $A$: The proof uses Theorem AI.1.8 and is similar to that of the corresponding theorem for Hilbert space operators.

Note that if $X$ is a Hilbert space, then the generalized adjoint is the usual Hilbert space adjoint.

We now discuss variational inequality and complementarity problem in semi-inner-product space under certain contractive type conditions on the operators.

Let $X$ be a real Banach space and $X^*$ its dual. The unit ball of $X$ is $\{x \in X: \|x\| \leq 1\}$ and its boundary $S = \{x \in X: \|x\| = 1\}$ is the unit sphere of $X$. $X$ is said to be smooth at $x \in S$ if there is a unique hyperplane of support at $x$, i.e., if there is only one continuous linear functional $q \in X^*$ with $\|g\| = 1$. $X$ is said to be smooth if it is smooth at every $x \in S$. Note that $X$ is smooth at $x \in S$ iff the norm in $X$ is Gateaux differentiable at that point. $X$ is said to be strongly smooth (ss for short) iff the norm in $X$ is Frechet differentiable. $X$ is said to be Uniformly convex (uc for short) if for each $\in$, $0 < \in \leq 2$, there is a $\delta = \delta(\in) > 0$ such that $x, y \in U$ and that $\|x + y\| \leq 2(1 - \delta(\in))$.

Let $E$ be a real vector space. $E$ is said to be a semi-inner-product space (sips for short) if there is a function

$[,]: E \times E \rightarrow R$ such that for $x, y, z \in E$ and $\lambda \in R$,

  (i) $[x + y, z] = [x, z] + [y, z]$,

     $[\lambda x, z] = \lambda[x, z]$,

  (ii) $[x, x] > 0$ for $x \neq 0$,

  (iii) $|[x, y]|^2 \leq [x, x] \, [y, y]$.

Note that a sip is a normed linear space (nls for short) with the norm: $\|x\| = [x, x]^{1/2}$. Further every nls can be made into a sips in infinitely many different ways and the sip is unique iff the space is smooth (Lumer [7]).

Observe that a sip is not linear in the second variable and it is so in the second variable iff it is an inner-product. However if $X$ is uc and smooth, then sip is homogeneous in the second variable (see Giles [3]).

Let $K$ be a closed convex subset of a sips $X$. If $T: K \rightarrow K$, then a variational inequality (VI for short) is stated as follows:

$$x \in K: \lfloor Tx, y - x \rfloor \geq 0 \quad \text{for all } y \in K.$$

If $K$ is a closed convex cone, then the polar or dual of $K$, denoted by $K^*$, is defined by

$$K^* = \{z \in X^*: [z, x] \geq 0 \quad \text{for all } x \in K\}.$$

If $K$ is a closed convex cone, then the complementarity problem (CP for short) is defined as follows:

Find $x \in X$ such that

$$x \in K, \; Tx \in K^* \text{ and } [Tx, x] = 0.$$

Observe that if $K$ is a closed convex cone, then (VI) and (CP) are equivalent.

We have

**Theorem AI.1.9**  Let $X$ be *uc* and *ss* and $K$ a nonempty closed convex subset of $X$.

Let $T: K \to K$ satisfy any one of the following conditions:

(i)  $\|Tx - Ty\| \le a \|x - y\| + b\|Tx - x\| + c\|Ty - y\|$

      where $-1 < a < 0$, $b \ge 0$, $c \ge 0$, $a + b + c = 0$,

(ii)  $\|Tx - Ty\| \le a_1 \|x - y\| + a_2\|x - Tx\|$

      $+ a_3\|y - Ty\| + a_4\|x - Ty\| + a_5 \|y - Tx\|$

where $-1 < a_1 < 0$, $a_2, a_3, a_4, a_5 \ge 0$, $\displaystyle\sum_{i=1}^{5} a_1 = 0$

Then there is a unique $y_0 \in K$ such that for all $x \in K$.

$$[Ty_0,\ z - y_0] \ge 0.$$

**Proof.**  Since $K$ is a nonempty closed convex subset of $X$ and $X$ is *uc*, for every $y \in K$ there is a unique $x \in K$ closest to $y - Ty$, that is,

$$\|x - y + Ty\| \le \|z - y + Ty\|$$

for every $z \in K$ (see Edelstein [1]). Let the correspondence $y \to x$ be denoted by $\theta$. Let $z \in K$ and let $0 \le \lambda \le 1$. Since $K$ is convex, $(1 - \lambda)\, x + \lambda z \in K$. Define a map

$h: [0, 1] \to R^+$ by

$$h(\lambda) = \|y - Ty - (1 - \lambda)x - \lambda z\|^2$$

Since $X$ is uc and ss, $h$ is a continuously differentiable function of $\lambda$ and

$$h'(\lambda) = 2[y - Ty - (1 - \lambda)x - \lambda z,\ x - z].$$

Since $x$ is the unique element closest to $y - Ty$ we must have $h'(0) \ge 0$. Therefore

$$[y,\ Ty - x,\ x - z] \ge 0 \quad \text{for all } z \in K \tag{1}$$

Let $y_1, y_2 \in K$ and $y_1 \ne y_2$ Let $q(y_1) = x_1$, $q(y_2) = x_2$.

It follows from (1) that

$$[y_1 - Ty_1 - \theta y_1,\ - \theta y_2 + \theta y_1] \ge 0 \tag{2}$$

and

$$[y_2 - Ty_1 - \theta y_2,\ - \theta y_2 - \theta y_1] \ge 0 \tag{3}$$

From (2) and (3) we get

$$[y_1 - Ty_1 - y_2 + Ty_2 - \theta y_1 + \theta y_2, \ \theta y_1 - \theta y_2] \geq 0$$

Therefore

$$\|\theta y_1 - \theta y_2\|^2 = [\theta y_1 - \theta y_1, \ \theta y_1 - \theta y_2]$$
$$= [y_1 - Ty_1 - y_2 + Ty_2, \ \theta y_1 - \theta y_2]$$
$$\leq \|y_1 - Ty_1 - y_2 + Ty_2\| \ \|\theta y_1 - \theta y_2\|$$

If $T$ satisfies condition (1) of Theorem 9, then

$$\|\theta y_1 - \theta y_2\| \leq (1 + a) \ \|y_1 - y_2\| + b\|Ty_1 - y_1\| + c\|Ty_2 - y_2\|$$

where $0 < 1 + a < 1$, $b \geq 0$, $c \geq 0$, $a + b + c = 1$.

If $T$ satisfies condition (ii) then

$$\|Tx - Ty\| \leq (1 + a_1) \ \|x - y\| + a_2 \ \|x - Tx\|$$
$$+ a_3\|y - Ty\| + a_4\|x - Ty\| + a_5\|y - Ty\|$$

where $0 < 1 + a_1 < 1$, $a_2, a_3, a_4, a_5 \geq 0$

$$(1 + a_1) + \sum_{i=1}^{4} a_i = 1$$

Therefore (see Reich [8], Gregus [4] and Ghosh [2] for case (i), Hardy and Rogers [5] for case (ii), Joshi and Bose [6] and Rhoades [9] for both the cases) $\theta$ has a unique fixed point, say $y_0$; hence $\theta(y_0) = y_0$. It now follows from (1) that for all $z \in K$,

$$[Ty_0, \ z - y_0] \geq 0$$

and this completes the proof.

**Theorem Al.1.10**   Let $X$ be Hilbert space and $K$ a closed convex cone and let the conditions of the previous theorem be satisfied. Then the CP has a unique solution, i.e., there is a unique $y_0 \in X$ such that

$$y_0 \in K, \ Ty_0 \ K^* \text{ and } (Ty_0, \ y_0) = 0.$$

**Proof.**   We have

$$(Ty_0, \ y_0 - z) \leq 0 \quad \text{for all } z \in K.$$

Since $0 \in K$, we get $(Ty_0, \ y_0) \leq 0$.

Since $K$ is a cone and $y_0 \in K$, it follows that $2y_0 \in K$ and hence

$$(Ty_0, \ y_0) \geq 0. \quad \text{Thus } (Ty_0, \ y_0) = 0$$

Further for all $z \in K$,

$$(Ty_0, z) \geq (Ty_0, y_0) = 0.$$

Thus $Ty_0 \in K^*$ and this completes the proof.

**Theorem AI.1.11**  Let $X$ be uc and ss and $K$ a nonempty closed convex subset of $X$.

Let $T: K \to K$ be nonexpansive. Then there exists some $y_0 \in K$ such that

$$\lfloor Ty_0 + y_0, y_0 \rfloor = 0.$$

**Proof.**  Proceeding as in Theorem 9, we obtain

$$\|\theta y_1 - \theta y_2\| \leq \|y_1 - Ty_1 - y_2 + Ty_2\|$$

If $T$ is nonexpansive, then

$$\|\theta y_1 - \theta y_2\| \leq \|Ty_1 - Ty_2\| + \|y_1 - y_2\|$$

$$\leq 2\|y_1 - y_2\|$$

Hence $\dfrac{\theta}{2}$ is nonexpansive and thus by Theorem 4.2.3, p. 98 of [6], $\dfrac{\theta}{2}$ has a fixed point, say $y_0$, i.e., $q(y_0) = 2y_0$. Hence it follows that

$$\lfloor Ty_0 + y_0, z - 2y_0 \rfloor \geq 0 \quad \text{for all } z \in K.$$

Since $0 \in K$, $\lfloor Ty_0 + y_0, y_0 \rfloor \leq 0$. Since $K$ is a cone and $y_0 \in K$, $3y_0 \in K$ and thus

$$\lfloor Ty_0 + y_0, y_0 \rfloor \geq 0.$$

Therefore $\lfloor Ty_0 + y_0, y_0 \rfloor = 0$ and this completes the proof.

## REFERENCES


1. M. Edelstein, On nearest points of sets in uniformly convex Banach spaces, J. London Math. Soc. 43 (1968) 375-77.

2. K. M. Ghosh, Fixed-point theorems, Pure Math Manuscript 3 (1984) 51-53.

3. J. R. Giles, Classes of semi-inner-product spaces, Trans. Amer. Math. Soc. 129 (1967), 436-46.

4. M. Gregus, A fixed-point theorem in Banach space, Boll. Un. Mat. Ital. 5 (17) − A (1980) 193-198.

5. G.E. Hardy and T. D. Rogers, A generalization of a fixed-point theorem of Reich, Canad, Math. Bull, 16 (1973) 201-206.

6. M. C. Joshi and R. K. Bose, some Topics in Nonlinear Functional Analysis, Wiley Eastern Ltd (1985).

7. G. Lumer, semi-inner-product spaces, Trans. Amer. Math. Soc. 100 (1961) 29-43.

8. S. Reich, some remarks concerning contraction mappings, Canad. Math. Bull. 14 (1971) 121-124, MR 45 # 1145.

9. B.E. Rhoades, A comparison of various definitions of contractive mappings, Trans. Amer. Math. Soc. 226 (1977) 257-289.

# Topological Vector Spaces and Hahn-Banach Theorems

The purpose of this chapter is to discuss some basic concepts of topological vector spaces and present Hahn-Banach Theorem and its generalizations.

## AII.1  TOPOLOGICAL VECTOR SPACES

A vector space $X$ over the field $K$ (R or C) together with a topology $\tau$ is called a linear topological space or topological vector space (in short tvs), if the vector space operations are continuous with respect to $\tau$, that is, if

    (i) $(x, y) \to x + y$ is continuous from $X \times X$ to $X$

and   (ii) $(\alpha, x) \to \tau x$ is continuous from $K \times X$ to $X$.

If the topology of $X$ is given by a metric, then $X$ is called a linear metric space. If the topology is given by a norm, then it is a normed linear space (nls in short).

Some simple properties are the following :

(i) If $G$ is an open set of $X$, then $x + G$ is also an open set for any $x \in X$ and $\lambda G$ is open for any $\lambda \in K$.

(ii) Every neighbourhood (nhd in short) of a point $x \in X$ (any open set containing the point $x$) has the form $x + V$ where $V$ is a nhd of the zero element of $X$.

By these properties, the topology $\tau$ is completely determined by any local base at 0. Thus, the local base at 0, simply local base $\beta$ of a tvs is a collection $\beta$ of neighbourhoods of 0 such that every nhd of 0 contains a member of $\beta$.

A set $A \subset X$ is said to be balanced if $\alpha A \subset A$ for every $\alpha \in K$ with $|\alpha| < 1$. A set $A$ is said to be bounded if for every nhd $V$ of 0, there exists a number $s > 0$ such that $A \subset tV$ for every $t > s$. A nhd $V$ of 0 is said to be symmetric if $V = -V$.

**Proposition AII.1.1**  If $V$ is a nhd of 0, then there is a symmetric nhd $U$ of 0 such that $U + U \subset V$.

**Proof.**   Since $0 + 0 = 0$ and the '+' is continuous, there exist neighbourhoods $W_1$ and $W_2$ such that

$$W_1 + W_2 \subset V$$

Let $U = W_1 \cap W_2 \cap (-W_1) \cap (-W_2)$. Then it is clear that $U$ is a symmetric neighbourhood of 0 with $U + U \subset V$.

In a tvs $X$, if every singleton set is closed, then the topology of $X$ turns out to be Hausdorff. In fact, we have the following separation property.

**Theorem AII.1.1**   Let $X$ be a tvs in which every singleton set is closed. Suppose $K$ and $C$ are subsets of $X$, $K$ is compact, $C$ is closed and $K \cap C = \phi$. Then there exists a neighbourhood $V$ of 0 such that

$$(K + V) \cap (C + V) = \phi.$$

**Proof.**   Here $K + V = \underset{x \in k}{\cup}\ x + V.$

Let $x \in K$. Since $K$ and $C$ are disjoint, $x \notin C$. Since $C$ is closed, there exists a symmetric nhd $V_x$ of 0 such that $(x + V_x + V_x + V_x) \cap C = \phi$. By the symmetry of $V_x$,

$$(x + V_x + V_x) \cap (C + V_x) = \phi$$

Since $K$ is compact, there exists finitely many points. $x_1 + x_2, \ldots x_n$ in $K$ such that $K \subset (x_1 + V_{x1}) \cup (x_2 + Vx_2) \cup \ldots \cup (x_n + Vx_n)$.

Put
$$V = V_{x_1} \cap \ldots \cap V_{x_n}$$

Now
$$K + V \subset \overset{n}{\underset{i=1}{\cup}} (x_i + V_{x_i} + V) \subset \overset{n}{\underset{i=1}{\cup}} (x_i + V_{x_i} + V_{x_i})$$

Since $x_i + V_{x_i} + V_{x_i} \cap C + V = \phi$, $(K + V) \cap (C + V) = \phi$

This completes the proof.

**Proposition AII.1.2**   If $V$ is a nhd of 0, there exists a balanced nhd $U$ of 0 such that $U \subset V$.

**Proof.**   Suppose $V$ is a nhd of 0. Since scalar multiplication is continuous, there is a $\delta > 0$ and there is a nhd $U$ of 0 such that $\alpha U \subset V$ whenever $|\alpha| < \delta$. Let $W$ be the union of all these sets $\alpha U$. Then $W$ is a nhd of 0, $W$ is balanced and $W \subset V$.

**Proposition AII.1.3**   If $V$ is convex nhd of 0, there exists a balanced convex nhd $U$ of 0 such that $U \subset V$.

**Proof.**   Let $V$ be a convex nhd of 0. Let $A = \underset{|\alpha|=1}{\cap}\ \alpha V$. By the above result, there, exists $W$, a balanced nhd of 0 such that $W \subset V$. Since $\alpha^{-1} W = W$

when $|\alpha| = 1$, $W \subset \alpha V$. Thus $W \subset A$. Then the interior $A^0$ of $A$ is a nhd of 0. Clearly $A^0 \subset V$. We observe that $A$ and hence $A^0$ are convex. If we prove that $A^0$ is balanced then $A^0$ is the required nhd of 0. For this, it is enough to show that $A$ is balanced. Choose $r$ and $\beta$ so that $0 \leq r \leq 1$, $|\beta| = 1$. Then

$$r\beta A = \bigcap_{\alpha} r\beta \alpha V = \bigcap_{|\alpha|=1} r\alpha V.$$

Since $\alpha V$ is a convex set that contains 0, we have $r\alpha U \subset \alpha U$. Thus $r\beta A \subset A$ which completes the proof.

From the Proposition AII.1.2, it follows that every topological vector space has a balanced local base.

**Theorem AII.1.2**   Suppose $V$ is neighbourhood of 0 in $X$.

(i) If $0 < r_1 < r_2 < \ldots\ldots$ and $r_n \to \infty$ as $n \to \infty$, then

$$X = \bigcup_{n=1}^{\infty} r_n V$$

(ii) If $\delta_1 > \delta > \ldots$ and $\delta n \to 0$ as $n \to \infty$, and if $V$ is bounded, then $\{\delta_n V : n = 1,2 \ldots\}$ is a local base for $X$.

**Proof.**   (i) Fix $x \in X$. Since $x \to \alpha x$ is continuous, the set of all $\alpha$ with $\alpha x \in V$ open, contains 0, hence contains $1/r_n$ for large $n$. Thus $\left(\dfrac{1}{r_n}\right) x \in V$ or $x \in r_n V$.

(ii) Let $U$ be a neighbourhood of 0 in $X$. If $V$ is bounded, there exists $s > 0$ such that $V \subset tV$ for all $t > s$. If $n$ is so large that $s\delta_n < 1$, it follows that $V \subset (1/\delta n)V$. Hence $V$ contains all but finitely many of the sets $\delta n \, V$.

A subset $A \subset X$ is said to be absorbing if every $x \in X$ lies in $tA$ for some $t > 0$. By the above theorem, every neighbourhood of 0 is observing.

The next theorem characterizes the boundedness of a subset $E$ of $X$ in terms of sequences.

**Theorem AII.1.3**   Let $E$ be a subset of a tvs. Then $E$ is bounded if and only if $\alpha_n x_n \to 0$ as $n \to \infty$ whenever $\{x_n\}$ is a sequence in $E$ and $\{\alpha_n\}$ is a sequence in the field $K$ with $\alpha n \to 0$ as $n \to \infty$.

**Proof.**   Suppose $E$ is bounded. Let $V$ be a balanced neighbourhood of 0 in $X$. Then $E \subset tV$ for some $t$. If $x_n \in E$ and $\alpha_n \to 0$, there exists $N$ such that $|\alpha_n| t < 1$ if $n > N$. Since $t^{-1} E \subset V$ and $V$ is balanced, $\alpha_n x_n \in V$ for all $n > N$. Therefore $\alpha_n x_n \to 0$.

If $E$ is not bounded, there is a neighbourhood $V$ of 0 and a sequence $r_n \to \infty$ such that $E \not\subset r_n V$ for all $n$. Choose $x_n \in E$ such that $x_n \not\subset r_n V$. Then $\dfrac{1}{r_n} x_n \notin V$ which implies that $\{r_n^{-1} x_n\}$ does not converge to 0.

## Locally Convex Spaces

A tvs $X$ is said to the locally convex if there is a local base consisting of convex neighbourhoods.

A semi norm on a vector space $X$ is a real valued function $P$ on $X$ such that

(i) $P(x + y) \leq p(x) + p(y)$

(ii) $P(\alpha x) \leq |\alpha|\, p(x)$ for all $x$, $y$ and all scalars $\alpha$.

Some simple consequences of this definition are:

(i) $P(0) = 0$

(ii) $|P(x) - P(y)| \leq P(x - y)$

(iii) $P(x) \geq 0$

(iv) $\{x : P(x) = 0\}$ is a subspace of $X$.

(v) The set $\{x : P(x) < 1\}$ is convex, balanced and absorbing.

Thus it is clear that if $P(x) \neq 0$ for $x \neq 0$ then $P$ is a norm on $X$.

A family $P$ of semi norms on $X$ is said to be separating if for each $x \neq 0$, there exists a $p \in P$ with $p(x) \neq 0$.

For every convex absorbing set $A$ in $X$, one defines the Minkowski functional $\mu_A$ as follows:

$$\mu_A(x) = \inf \left\{ t > 0 : \frac{1}{t}\, x \in A \right\}, \quad x \in X.$$ We note that since $A$ is absorbing, $0 \in A$ and $\mu_A(x) < \infty$ for all $x \in X$.

**Proportion AII.1.4**  If $A$ is convex absorbing and balanced, then $\mu_A$ is a semi norm on $X$.

**Proof.**  For $x \in X$ let $H_A(x) = \left\{ t > 0 : \frac{1}{t}\, x \in A \right\}$

Let $t \in H_A(x)$ and $s > t$. Since $0 \in A$ and $A$ is convex, it follows that $s \in H_A(x)$.

Let $\mu_A(x) < s,\ \mu_A(y) < t, \quad u = s + t$

Then $s^{-1}x \in A$, $t^{-1}y \in A$. Since $A$ is convex.

$$u^{-1}(x + y) = \frac{s}{u}(s^{-1}x) + \left(\frac{t}{u}\right)(t^{-1}y) \in A.$$

Hence $\mu_A(x + y) \le u$. Thus $\mu_A(x + y) \le \mu_A(x) + \mu_A(y)$.

It is easy to see that $\mu_A(\alpha x) = |\alpha|\,\mu_A(x)$ for all $x \in X$ and $\alpha \in K$.

**Theoremm AII.1.4**   Let $\beta$ be a convex balanced local base in a tvs $X$. Let $\mu_v$ denote Minkowski functional of $V \in \beta$. Then $\{\mu_V : V \in \beta\}$ forms a separating family of continuous semi norms.

**Proof.**   Since $V \in \beta$ is convex balanced and absorbing, by the above preposition. $\mu_V$ is a semi norm. Let $x \ne 0 \in X$. Then there exists $V \in \beta$ such that $x \in V$. Then $\mu_V(x) \ge 1$. Thus $\{\mu_V\}$ is a separating family. If $x \in V$, since $V$ is open, $tx \in V$ for some $t > 1$. Hence $\mu_V(x) < 1$ for all $x \in V$. Let $s > 0$. If $x-y \in sV$, then

$$|\mu_V(X) - \mu_V(y)| \le \mu_V(x - y) < s$$

That is $\mu_V$ is continuous.

## Other Types of Topological Vector Spaces

A tvs $X$ is

(i) locally bounded if 0 has a bounded neighbourhood.

(ii) locally compact if 0 has a neighbourhood whose closure is compact.

(iii) metrizable if $\tau$ is compatible with a metric $d$.

(iv) $F$-space if $\tau$ induced by a complete invariant metric $d$ (in the sense that

$$d(x + y, y + z) = d(x, y)\ \text{for}\ x, y, z \in X).$$

(v) Frechet space it is a locally convex $F$-space.

(vi) Normable if a norm exists on $X$ such that the metric induced by the norm is compatible with $\tau$.

**Theoremm AII.1.5**   A Hausdoff topological vector space $X$ is normable if and only if its origin has a convex bounded neighbourhood, that is, if and only if $X$ is locally convex and locally bounded.

**Proof.**   Let $V$ be convex bounded neighbourhood of 0. Then $V$ contains a convex balanced neighbourhood $V$ of 0. Define $\|x\| = \mu(x)$, $x \in X$, where $\mu$ is the Minkowski functional of $V$.

By (ii) of Theorem AII.1.2, $\{rv : r > 0\}$ forms a local base for the topology of $X$. if $x \neq 0$, then $x \notin rV$ for some $r > 0$. Then $\|x\| \geq r$. Thus $\|.\|$ defines a norm on $X$. We further note that $\{x : \|x\| < r\} = rV$ for all $r > 0$. Therefore the topology induced by the norm coincides with the topology of $X$. The converse part is obvious.

## AII.2  HAHN-BANACH THEOREM AND GENERALIZATIONS

In this section we shall present the famous Hahn-Banach Theorem. This theorem is one of the most fundamental theorems in Functional analysis and is due to Hahn and Banach independently. It yields the existence of non-trivial continuous linear functionals on a normed linear space, a basic result necessary for the development of a large portion of Functional Analysis. Moreover it has wide applications and is an indispensable tool in the proofs of many important theorems in Analysis. There are several forms of this famous theorem and the one, which will be proved first, is known as the extended form of Hahn-Banach Theorem. We shall then present several generalizations of this famous theorem.

**Theorem AII.2.1**  Let $E$ be a real linear space and let $M$ be a linear subspace of $E$. Suppose that $p$ is a sublinear functional defined on $E$ and that $f$ is a linear functional defined on $M$ such that $f(x) \leq p(x)$ for every $x \in M$. Then there is a linear functional $g$ defined on $E$ such that $g$ is an extension of $f$ (i.e., $g(x) = f(x)$ for all $x \in M$) and $g(x) \leq p(x)$ for all $x \in E$.

**Proof.**  Let $\Im$ denote the set of all real functions, $h$, such that $h$ is linear, dom $h$ is a linear subspace of $E$, $h$ is an extension of $f$ and $h(x) \leq p(x)$ for all $x \in$ dom $h$. Since $f \in \Im$, $\Im \neq \Phi$. We write $f \subset h$ to denote that $h$ is an extension of $f$ (i.e., dom $f \subset$ dom $h$ and $f(x) = h(x)$ for $x \in$ dom $f$). Notice that $\Im$ is a partially ordered set with respect to the partial order $\subset$. Let $\mathbb{C}$ be any chain in $\Im$ and let $h = \cup C$. Then $h \in \Im$. Therefore from Zorn's lemma it follows that $\Im$ has a maximal element, say $g$. We complete the proof by showing that dom $g = E$. Assume that this is false i.e., let dom $g = G \subset E$. Let $y$ be any element in $E \cap G^C$. Define

$$H = \{x + \alpha y : x \in G,\ \alpha \in R\}$$

Clearly $H$ is a linear subspace of $E$ and $G \subsetneq H$. Let $c$ be a fixed, but arbitrary, real number. Define $h$ on $H$ by

$$h(x + \alpha y) = g(x) + \alpha c$$

Now if $x_1 + \alpha_1 y = x_2 + \alpha_2 y$, where $x_1, x_2 \in R$, then $(\alpha_1 - \alpha_2)y = x_2 - x_1 \in G$ so that $\alpha_1 = \alpha_2$ and $x_1 = x_2$. Hence $h$ is well defined. Clearly $h$ is linear and $g \subset h$. We now claim that a '$c$' can be selected so that $h(x) \leq p(x)$ for all $x \in^{\neq} H$; then we have $h \in \mathfrak{I}$, which contradicts the maximality of $g$ and completes the proof of the theorem. Therefore we need only to establish our claim by showing that $c$ can be so selected.

Thus our requirement is that $g(x) + \alpha c = h(x + \alpha y) \leq p(x + \alpha y)$ for all $x \in G$, $\alpha \in R$. Since $g$ is linear and $p$ is sublinear, it is equivalent to requiring that

$$g\left(\frac{x}{\alpha}\right) + c \leq p\left(\frac{x}{\alpha} + y\right) \text{ for } x \in G \text{ and } \alpha > 0,$$

and

$$g\left(\frac{x}{\alpha}\right) + c \geq - p\left(-\frac{x}{\alpha} - y\right) \text{ for } x \in G \text{ and } \alpha < 0,$$

Therefore it suffices to have

$$g(u) - f(u - y) \leq c \leq - g(v) + p(v + y) \text{ for } u, v \in G$$

But we do have

$$g(u) + f(v) = g(u + v) \leq p(u - y) < p(u - y) + p(v + y)$$

for all $u, v \in G$. Write

$$a = \sup\{g(u) - p(u - y) : v \in G\}$$

and

$$b = \inf\{-g(v) + p(v + y) : v \in G\}$$

It is clear that $a \leq b$. Now any real number $c$ such that $a \leq c \leq b$ satisfies our requirements.

**Theorem AII.2.2**  Let $E$ be a real normed linear space and let $M$ be a linear subspace of $E$. If $f \in M^*$ then there is $g \in E^*$ such that $f \subset g$ and $\|g\| = \|f\|$.

**Proof.**  Define $p$ on $E$ by

$$p(x) = \|f\| \, \|x\|$$

Then clearly $p$ is a sublinear functional on $E$ and $f(x) \leq |f(x)| \leq p(x)$ for all $x \in M$. Therefore it follows from Theorem (1.1) that there exists a linear functional $g$ on $E$ such that $f \subset g$ and $g(x) \leq p(x)$ for all $x \in E$. Clearly $g \in E^*$ and $\|g\| \leq \|f\|$. Also

$$g = \sup\ \{|g(x)| : x \in E,\ \|x\| = 1\}$$
$$\geq \sup\ \{|g(x)| : x \in M,\ \|x\| = 1\}$$
$$= \sup\ \{|f(x)| : x \in M,\ \|x\| = 1\}$$
$$= \|f\|.$$

Therefore $\|g\| = \|f\|$ and this completes the proof.

**Theorem AII.2.3**   Let $E$ be a complex linear space and let $M$ be a linear subspace of $E$. Suppose $q$ is a seminorm on $E$ and $f$ is a linear functional defined on $M$ such that $|f(x)| \leq q(x)$ for all $x \in M$. Then there is a linear functional $g$ on $E$ such that $f \subset g$ and $|g(x)| \leq q(x)$ for all $x \in E$.

**Proof.**   For each $x \in M$, write $f(x) = f_1(x) + if_2(x)$.

An easy computation shows that $f_1$ and $f_2$ are real linear functionals on $M$. It is also obvious that for $j = 1, 2$,

$$|f_j(x)| \leq |f(x)| \leq q(x)$$

for all $x \in M$. Now regard $E$ and $M$ as real linear spaces and apply theorem (1.1) to obtain a real linear functional $g_1$ on $E$, such that $f_1 \subset g_1$ and

$$|g_1(x)| \leq |q(x)| \qquad \text{for all } x \in E.$$

Now define $g$ on $E$ by the rule

$$g(x) = g_1(x) - i\ g_1(ix),$$

It is easy to see that $g$ is a complex linear functional on $E$. Further for $x \in E$,

$$g_1(ix) + if_2(ix) = f_1(ix) + if_2(ix)$$
$$= f(ix) = if(x)$$
$$= -f_2(x) + if_1(x)$$
$$= -f_2(x) + ig_1(x)$$

so that $g_1(ix) = -f_2(x)$ and therefore

$$g(x) = g_1(x) - ig_1(ix) = f_1(x) + if_2(x) = f(x)$$

Thus $f \subset g$. It remains only to show that

$$|g(x)| \leq q(x) \qquad \text{for all } x \in E.$$

Let $x \in E$ be arbitrary. Write $g(x) = re^{i\theta}$, $r \geq 0$, $\theta \in R$. Then

$$|g(x)| = r = e^{-i\theta}\ g(x) = g(e^{-i\theta}\ x)$$
$$= g_1(e^{-i\theta}x) \leq q(e^{-i\theta}x)$$
$$= |e^{-i\theta}|\ q(x) = q(x)$$

This completes the proof.

**Theorem AII.2.4**   Let $E$ be a complex normed linear space and let $M$ be a linear subspace of $E$. If $f \in M^*$, then there exists $f \in E^*$, such that $f \in g$ and $\|g\| = \|f\|$. This can be obtained by arguing as in Theorem (AII.2.2) and using Theorem (AII.2.3). We now present some results, which are applications of Hahan-Banach Theorem.

**Theorem AII.2.5**   Let $E$ be a normed linear space and let $S$ be a linear subspace of $E$. Suppose that $z \in E$ and dist $(z, S) = d > 0$. Then there exists $g \in E^*$ such that $g(S) = \{0\}$, $g(z) = d$ and $\|g\| = 1$.

**Proof.**   Let

$$M = \{x + \alpha z : z \in S, \alpha \in K\}.$$

Then clearly $M$ is a linear subspace of $E$. Define $f$ on $M$ by

$$f(x + \alpha z) = \alpha d$$

Clearly $f$ is a well-defined linear functional on $M$. Also $f(S) = \{0\}$ and $f(z) = d$. Further

$$\|f\| = \sup\left\{ \frac{|f(x + \alpha z)|}{\|x + \alpha z\|} : x + \alpha z \in M,\ \| x + \alpha z\| \neq 0 \right\}$$

$$= \sup\left\{ \frac{|\alpha d|}{\|x + \alpha z\|} : x + \alpha z \in M,\ \|x + \alpha z\| \neq 0 \right\}$$

$$= \sup\left\{ \frac{d}{\|-y + z\|} : y \in S \right\}$$

$$= \frac{\alpha}{d} = 1.$$

Thus $f \in M^*$. Now by applying Theorem (AII.2.2) or (AII.2.4), as the case may be, we obtain a functional $g \in E^*$ which satisfies the requirements of the theorem.

**Corollary AII.2.1**   Let $E$ be a normed linear space and let $z$ be a non-zero vector in $E$. Then there exists a functional $g \in E^*$ such that $g(z) = \|z\|$ and $\|g\| = 1$.

**Proof.**   Take $S = \{0\}$ in Theorem (AII.2.5).

This answers the question raised in the previous section that if $E$ is a nonzero normed linear space, then there exists a nonzero element in $E^*$.

**Discussion AII.2.1**   If $E$ is a normed linear space, then there exists a natural mapping from $E$ into $E^{**}$. Each element $x \in E$ gives rise to a functional $F_x$ in $E^{**}$ defined by

$$F_x(f) = f(x)$$

for $f \in E^*$. We denote the natural mapping $x \to F_x$ from $E$ into $E^{**}$ by $\pi$. A simple computation shows that $F_x$ is a linear functional on $E^*$. Furthermore

$$\|f\| = \sup\{|F_x(f)| \quad : \|f\| \leq 1\}$$
$$= \sup\{|f(x)| \quad : \|f\| \leq 1\}$$
$$\leq \sup\{\|f\| \, \|(x)\| \quad : \|f\| \leq 1\}$$
$$\leq \|x\|$$

Thus $F_x \in E^{**}$ and $\pi$ is well-defined. Also the mapping $\pi$ is linear and since

$$\|\pi(x)\| = \|F_x\| \leq \|x\|,$$

it follows that $\pi$ is a bounded linear transformation from $E$ into $E^{**}$ with $\|\pi\| \leq 1$. Several questions now arise:

**(1)** Is $\pi$ $1-1$ ? (2) Does it preserve norms ? (3) Is it onto $E^{**}$ ? We are able to answer (1) and (2) with the help of Hahn-Banach Theorem, infact with the help of Corollary AII.2.1, and this has been presented in the following theorem.

**Theorem AII.2.6**   Let $E$ be a normed linear space and let $\pi$ be the natural mapping from $E$ into $E^{**}$ defined in Discussion AII.2.1. Then $\pi$ is a (bounded) norm-preserving linear transformation from $E$ into $E^{**}$ and $\pi$ is $1-1$ .

**Proof.**   Observe that $\pi$ is a bounded linear transformation from $E$ into $E^{**}$ with the norm $\|\pi\| \leq 1$. Let $x$ be any nonzero element of $E$. Then by Corollary (AII.2.1), there exists an element $g \in E^*$ such that $\|g\|$ and $g(x) = \|x\|$. Now

$$\|x\| = g(x)$$
$$\leq \sup\{|f(x)| : f \in E^{**}, \|f\| \leq 1\}$$
$$= \sup\{|F_x(f)| : f \in E^{**}, \|f\| \leq 1\}$$
$$= \|F_x\| = \|\pi(x)\| \leq \|x\|$$

Thus $\|\pi(x)\| = \|x\|$. Clearly $\|\pi(0)\| = 0 = \|0\|$ and hence $\pi$ preserves norms. Consequently, if $x, y \in E$ and $x \neq y$, then $\|\pi(x) - \pi(y)\| = \|\pi(x-y)\| = \|x-y\| \neq 0$; so $\pi(x) \neq \pi(y)$ and hence $\pi$ is $1-1$. This completes the proof.

**Remark AII.2.1**   If $E$ is a normed linear space, it follows from Theorem AII.2.6 that the natural mapping $\pi$ is an isometric isomorphism from $E$ into $E^{**}$. This allows us to identify $E$ with the subspace $\pi(E)$ of $E^{**}$ and with respect to this identification $E$ is regarded as a subspace of $E^{**}$. It shall be noted that $\pi$ need not be onto. A normed linear space $E$ is said to be reflexive if $\pi$ is onto, i.e., if $\pi(E) = E^{**}$; in other words $E$ is reflexive if $E$ and $E^{**}$ are indistinguishable as normed linear spaces.

Let $X$ be a vector space over the field $R$ of real numbers. A nonempty subset $E$ of $X$ is called a semi-space (or a cone) if

(i) $x, y \in E$ imply $x + y \in E$,

(ii) $x \in E$ and $\lambda \geq 0$ imply $\lambda x \in E$.

A functional $p$ defined on a cone $E$ is said to be sublinear on $E$ if for $x, y \in E$, $\lambda \geq 0$,

(i) $p(x + y) \leq p(x) + p(y)$,

(ii) $p(\lambda x) = \lambda p(x)$

A functional $q$ is said to be super linear on $E$ if $- q$ is sublinear on it.

**Theorem AII.2.7**   Let $E$ be a cone in a vector space $X$ over $R.$. Let $p$ be a sublinear functional on $X$ and let $q$ be a superlinar functional on $E$ with the property that

$$q(x) \leq p(x) \text{ for all } x \in E.$$

Then there exists a linear functional $f$ on $X$ such that

$$f(x) \leq p(x) \text{ for } x \in X,$$

$$f(x) \geq q(x) \text{ for } x \in E.$$

**Remark AII.2.2**   In the above theorem if $E$ is a subspace of $X$ and $q$ is a linear functional on $E$, then this reduces to the usual extended form of Hahn-Banach Theorem.

**Proof. of Theorem AII.2.7**   Let $r$ be the functional defined on $X$ by

$$r(x) = \inf \{p(x + y) - q(y) : y \in E\}$$

We have

$$p(x + y) + p(-x) \geq p(y)$$

and therefore $r(x)$ is finite. It is easy to verify that

(i) $r$ is a sublinear functional on $X$,

(ii) $r(x) \leq p(x) \; \forall x \in X,$ $\hspace{6cm}$ (1)

(iii) $r(-x) \leq -q(x) \; \forall x \in E,$ $\hspace{5.5cm}$ (2)

By the usual extended form of Hahn-Banach theorem, there exists a linear functional $f$ on $X$ with

$$f(x) \leq p(x) \hspace{3cm} \forall x \in X, \hspace{3cm} (3)$$

It now follows from (1) and (3) that

$$f(x) \le p(x) \qquad \forall\ x \in X,$$

and from (2) and (3) that

$$f(x) \ge q(x) \qquad \forall x \in E.$$

This completes the proof.

# Bibliography


1. R. P. Agnew, Linear functionals satisfying prescribed conditions, Duke Math J 4 (1938) 55-77.

2. R. P. Agnew and A. P. Morse, Extensions of linear functionals with application to limits, integrals, measures and densities, Annals Math. (2) 39 (1938) 20-30.

3. F.F. Bonsall, the decomposition of continuous linear functionals into nonnegative components, Proc. Univ. Durham Philos Soc XII (1957) (2) 6-11.

4. F. F. Bonsall and A. W. Goldie, Algebras which represent their linear functionals, Proc. Camb. Philos Soc. 49 (1953) 1-14.

5. F. F. Bonsall, Sublinear functionals and ideals in partially ordered vector spaces, Proc. London Math Soc. 3 (4) (1954) 402-18.

6. H. F. Bohnenblust and A. Sonczyk, Extension of linear functionals on complex linear spaces, Bull. Amer. Math. Soc. 44 (1938) 91-93.

7. F. Munchen Behringer, On Karamardian's theorem about lower semi-continuous strictly quasiconvex functions, ZOR, Soc. A, 23 (1970) 17-48.

8. A. Ben-Tal, On generalized means and generalized convex functions, JOTA 21 (1977) 1-13.

9. B. Bereanu, On the composition of convex functions, Rev. Roum. Math Pures Appl. 19 (1969) 1078-84.

10. M. S. Bazaras, T. J. Goode and M. Z. Nashed, A nonlinear complementarity problem in mathematical programming in Banach space, Proc. Amer. Math. Soc. 35 (1972), 165-170.

11. F. E. Browder, Nonlinear variational inequalities and maximal monotone mappings in Banach spaces, Math. Annalen, 176 (1968), 89-113.

12. _______________ , Existence and approximation of solutions of nonlinear variational inequalities, Proc. Natl. Acad. Sci. U. S. 56 (1966a), 1080-1086.

13. _______________, On the unification of calculus of variation and the theory of monotone nonlinear operators in Banach spaces, ibid (1966b), 419-425.

14. _______________, Nonlinear monotone operators and convex sets in Banach spaces, Bull. Amer. Math. Soc. 71(1965a) 780-58.

15. _______________, On a theorem of Beurling and Livingston, Canad. J. Math. 17 (1965b), 367-372.

16. _______________, Variational boundary value problems for quasilinear elliptic equations of arbitrary order, Proc. Natl. Acad. Sci., U. S. 50 (1963), 31-37.

17. B. D. Craven and B. Mond. Complementarity over arbitrary cone, Z. O. R. (1977).

18. R. Chandrasekharan, A special case of the complementarity pivot problem, Opsearch, 7 (1970) 263-268.

19. R. W. Cottle and G. B. Dantzig, Complementarity pivot theory of mathematical programming, Linear Algebra and its Appl. 1 (1968), 108-125.

20. O. D. Donato and G. Maier, Mathematical-programming methods for the inelastic analysis of reinforced concrete frames allowing for limited rotation capacity, Int. J. Numer. Methods Eng. 4 (1972) 307-329.

21. A. T. Dash and S. Nanda, A complementarity problem in mathematical programming in Banach space, J. Math. Analysis Appl. 98 (1984).

22. B. C. Eaves, On the basic theorem of completemtarity, Math. Programming 1 (1971) 68-75.

23. B. Garcia, Some classes of matrices in Linear complementarity problem Math. Programming 5 (1973), 299-310.

24. Hartman and G. Stampacchia, On some nonlinear elliptic differential functional equations, Acta Math. 125 (1956) 271-310.

25. G. J. Habetler and A. L. Price, Existence theory for generalized nonlinear complementarity problem, J. Optim. Theory Appl. 7 (1971) 223-239.

26. G. J. Habetler and M. M. Kostreva, On a direct algorithm for nonlinear complementarity problems SIAM Journ. Control and Optim. 16(3) (1978) 504-511.

27. S. Karamaradian, The nonlinear complementarity problem with applications I, II, J. Optm. Theory Appl. 4 (1969) 87-98, 167-181.

28. _______________, Generalized complementarity problem, ibid 8 (1971) 161-168.

29. J. L. Lions and G. Stampacchia, Variational inequalilities, Comm. Pure Appl. Math. 20 (1967) 493-519.

30. C. E. Lemke, Bimatrix equilibrium points and mathematical programming, Management Sci. Ser. A, 11 (1965) 168-169.

31. ________________, Recent results on complementarity problems, in Nonlinear programming (j. B. Rosen, O.L. Mangasarian and K. Ritter, eds.), Academic New York (1970).

32. C. E. Lemke and J. T. Howson Jr., Equilibrium points of bimatrix games, SIAM J. Appl. Math. 12 (1964), 413-423.

33. B. Mond. On the complex complementarity problem, Bull. Austral. Math. Soc. 9 (1973) 249-257.

34. O. G. Mancino and G. G. Stampacchia, Convex programming and variational inequalities, J. Optm. Theory Appl. 9(1) (1972), 3-23.

35. U. Mosco. Convergence of convex sets and solutions of variational inequalities, Adv. Math. 3 (1966), 510-585.

36. ________________, A remark on a theorem of F. E. Browder, J. Math. Analysis Appl. 20 (1967a), 90-93.

37. ________________, Approximation of the solution of some variational inequalities, Ann Scuolo Normals sup. Pisa 21 (1967b), 373-934; 765.

38. K. G. Murty, On a characterization of P-matrices, SIAM J. Appl. Math. 20(3) (1971), 378-384.

39. ________________, On the number of solutions of the complementarity problem and spanning properties of complementary cones, Linear Algebra and its Appl. 5 (1972), 65-108.

40. ________________, Note on a Bard-type scheme for solving the complementarity problem, Opsearch 11, 2-3 (1974), 123-130.

41. Sribatsa Nanda and Sudarsan Nanda, A complex nonlinear complementarity problem, Bull. Austral. Math. Soc. 19 (1978) 437-444.

42. ________________, On stationary points and the complementarity problem, ibid 20 (1979a) 77-86.

43. ________________, A nonlinear complementarity problem in mathematical programming in Hilbert space, ibid 20 (1979b) 233-236.

44. ________________, A nonlinear complementarity problem in Banach space ibid 21 (1980) 351-356.

45. Sudarsan Nanda A note on a theorem on a nonlinear complementarity problem ibid 27 (19783) 161-163.

46. J. Parida and B. Sahoo, On the complex nonlinear complementarity problem, ibid 14 (1976a), 129-136.

47. ________________, Existence theory for the complex nonlinear complementarity problem, ibid 14 (1976b), 417-435.

48. A. Ravindran, A computer routine for quadratic and linear programming problems [H], Commun. ACM 15 (9) (1972) 818-820.

49. ________________, A comparison of the primal-simplex and complementary pivot methods for linear programming, Naval Res. Logistics Q. 20(1) (1973) 96-100.

50. H. Scarf, The approximation of fixed points of a continuous map. SIAM J. Appl. Math. 15 (1967), 328-343.

51. R. Saigal, On the class of complementary cones and Lemkes' algorithm, SIAM Journ. Appl. Math. 23 (1972) 46-60.

52. ________________, A note on a special linear complementarity problem, Opsearch 7 (3) (1970) 175-183.

53. S. Simons, Variational inequalities via the Hann-Banach theorem, Arch. Math. 31 (1978) 482-490.

54. R. J. Silverman, Invariant linear functions, Trans. Amer. Math. Soc. 81 (1956) 411-24.

55. G. V. Smith, The Hahn-Banach theorem for modules, Proc. London Math. Soc. 3 (17) (1968) 72-90.

56. F. A. Valentine, Convex sets, MeGraw Hill, New York (1964).

57. F. B. Wright, Generalized means, Trans Amer Math. Soc. 98 (1961) 187-203.

58. ________________, Corrections to generalized means, ibid 100 (1961) 370.

59. L. T. Watson, Solving nonlinear complementarity problem by Homotopy methods, SIAM J. Control and Optm. 17 (1) (1979) 36.

60. T. Weir, Generalized convexity and duality in mathematical Programming, Ph. D. Thesis, La Trobe University, Bundoora, Australia (1982).

# Index

# About the Author

Sudarsan Nanda is at present the Professor of Eminence and Research Chair at KIIT University, Bhubaneswar.

He is a Ph.D and D.Sc. in Mathematics and was a Professor at IIT Kharagpur. He was Vice-Chancellor, North Orissa University, Professor at Utkal University, Assistant Professor at IIT Kharagpur and Lecture at NIT Raurkela. Professor Nanda was a Visiting Professor at the Universities of Guelph, Canada, Pisa, and Milan in Italy, University of Central Folorida, USA and Chinese University of Hongkong. He was an Associate member of ICTP, Trieste, Italy, visited University of Kaiserlautern, Germany under exchange programme and several Universities of Europe and USA.

S. Nanda is a Fellow of Institute of Mathematics and Applications, U.K. (FIMA), Forum D'e Analysts, Member of the Editorial Board of the Journal of Fuzzy Mathematics and several other Journals on Mathematics. He is a reviewer of Mathematical reviews and several journals of international repute. Prof. Nanda received the distinguished teacher award during 2002, and received Fulbright travel grant for visiting USA, during 2000.

S. Nanda has published around 200 research papers in the journals of International repute, around 50 papers in national and international conferences, guided 18 Ph.D. students and authored and co-authored 16 books and many popular articles relating to Mathematics and Higher education.

S. Nanda's research interest is in Sequence Spaces, Nonlinear Functional Analysis, Optimization, Topology, Discrete and Fuzzy Mathematics. He obtained existence theorem on variational inequality and symmetric duality under most general conditions, obtained and studied new sequence spaces and introduced vector spaces and algebras for fuzzy sets.

S. Nanda's research work has been widely cited; he has delivered invited talks in many international and national conferences in India, Europe, and USA and has been members in many academic bodies in several Universities and Institutions.